TK 3226 .D36 2006

De La Rosa, Francisco C.

Harmonics and power systems

NEW ENGLAND INSTITUTE OF TECHNOLOGY
LIBRARY

HARMONICS AND POWER SYSTEMS

The ELECTRIC POWER ENGINEERING Series
Series Editor Leo L. Grigsby

Published Titles

Electric Drives
Ion Boldea and Syed Nasar

**Linear Synchronous Motors:
Transportation and Automation Systems**
Jacek Gieras and Jerry Piech

**Electromechanical Systems, Electric Machines,
and Applied Mechatronics**
Sergey E. Lyshevski

Electrical Energy Systems
Mohamed E. El-Hawary

Distribution System Modeling and Analysis
William H. Kersting

The Induction Machine Handbook
Ion Boldea and Syed Nasar

Power Quality
C. Sankaran

Power System Operations and Electricity Markets
Fred I. Denny and David E. Dismukes

Computational Methods for Electric Power Systems
Mariesa Crow

Electric Power Substations Engineering
John D. McDonald

Electric Power Transformer Engineering
James H. Harlow

Electric Power Distribution Handbook
Tom Short

Synchronous Generators
Ion Boldea

Variable Speed Generators
Ion Boldea

Harmonics and Power Systems
Francisco C. De La Rosa

HARMONICS AND POWER SYSTEMS

Francisco C. De La Rosa

Distribution Control Systems, Inc.
Hazelwood, Missouri, U.S.A.

Taylor & Francis
Taylor & Francis Group
Boca Raton London New York

A CRC title, part of the Taylor & Francis imprint, a member of the
Taylor & Francis Group, the academic division of T&F Informa plc.

Published in 2006 by
CRC Press
Taylor & Francis Group
6000 Broken Sound Parkway NW, Suite 300
Boca Raton, FL 33487-2742

© 2006 by Taylor & Francis Group, LLC
CRC Press is an imprint of Taylor & Francis Group

No claim to original U.S. Government works
Printed in the United States of America on acid-free paper
10 9 8 7 6 5 4 3 2 1

International Standard Book Number-10: 0-8493-3016-5 (Hardcover)
International Standard Book Number-13: 978-0-8493-3016-2 (Hardcover)
Library of Congress Card Number 2005046730

This book contains information obtained from authentic and highly regarded sources. Reprinted material is quoted with permission, and sources are indicated. A wide variety of references are listed. Reasonable efforts have been made to publish reliable data and information, but the author and the publisher cannot assume responsibility for the validity of all materials or for the consequences of their use.

No part of this book may be reprinted, reproduced, transmitted, or utilized in any form by any electronic, mechanical, or other means, now known or hereafter invented, including photocopying, microfilming, and recording, or in any information storage or retrieval system, without written permission from the publishers.

For permission to photocopy or use material electronically from this work, please access www.copyright.com (http://www.copyright.com/) or contact the Copyright Clearance Center, Inc. (CCC) 222 Rosewood Drive, Danvers, MA 01923, 978-750-8400. CCC is a not-for-profit organization that provides licenses and registration for a variety of users. For organizations that have been granted a photocopy license by the CCC, a separate system of payment has been arranged.

Trademark Notice: Product or corporate names may be trademarks or registered trademarks, and are used only for identification and explanation without intent to infringe.

Library of Congress Cataloging-in-Publication Data

De la Rosa, Francisco.
    Harmonics and power systems / by Francisco De la Rosa.
        p. cm.
    Includes bibliographical references and index.
    ISBN 0-8493-3016-5
    1. Electric power systems. 2. Harmonics (Electric waves) I. Title.

TK3226.D36 2006
621.31'91--dc22   2005046730

Visit the Taylor & Francis Web site at
http://www.taylorandfrancis.com

Taylor & Francis Group
is the Academic Division of Informa plc.

and the CRC Press Web site at
http://www.crcpress.com

To the memory of my father and brother
To my beloved mother, wife, and son

Preface

This book seeks to provide a comprehensive reference on harmonic current generation, propagation, and control in electrical power networks. Harmonic waveform distortion is one of the most important issues that the electric industry faces today due to the substantial volume of electric power that is converted from alternating current (AC) to other forms of electricity required in multiple applications. It is also a topic of much discussion in technical working groups that issue recommendations and standards for waveform distortion limits. Equipment manufacturers and electric utilities strive to find the right conditions to design and operate power apparatuses that can reliably operate in harmonic environments and, at the same time, meet harmonic emission levels within recommended values.

This book provides a compilation of the most important aspects on harmonics in a way that I consider adequate for the reader to better understand the subject matter. An introductory description on the definition of harmonics along with analytical expressions for electrical parameters under nonsinusoidal situations is provided in Chapter 1 as a convenient introductory chapter. This is followed in Chapter 2 by descriptions of the different sources of harmonics that have become concerns for the electric industry.

Industrial facilities are by far the major producers of harmonic currents. Most industrial processes involve one form or another of power conversion to run processes that use large direct current (DC) motors or variable frequency drives. Others feed large electric furnaces, electric welders, or battery chargers, which are formidable generators of harmonic currents. How harmonic current producers have spread from industrial to commercial and residential facilities — mostly as a result of the proliferation of personal computers and entertaining devices that require rectified power — is described. Additionally, the use of energy-saving devices, such as electronic ballasts in commercial lighting and interruptible power supplies that provide voltage support during power interruptions, makes the problem even larger.

As this takes place, standards bodies struggle to adapt present regulations on harmonics to levels more in line with realistic scenarios and to avoid compromising the reliable operation of equipment at utilities and customer locations. The most important and widely used industry standards to control harmonic distortion levels are described in Chapter 3.

The effects of harmonics are thoroughly documented in technical literature. They range from accelerated equipment aging to abnormal operation of sensitive processes or protective devices. Chapter 4 makes an effort to summarize the most relevant effects of harmonics in different situations that equally affect residential, commercial, and industrial customers. A particular effort is devoted to illustrating the effects of harmonics in electrical machines related to pulsating torques that can drive machines into excessive shaft vibration.

Given the extensive distribution of harmonic sources in the electrical network, monitoring harmonic distortion at the interface between customer and supplier has become essential. Additionally, the dynamics of industrial loads require the characterization of harmonic distortion levels over extended periods. Chapter 5 summarizes the most relevant aspects and industry recommendations to take into account when deciding to undertake the task of characterizing harmonic levels at a given facility.

One of the most effective methods to mitigate the effect of harmonics is the use of passive filters. Chapter 6 provides a detailed description of their operation principle and design. Single-tuned and high-pass filters are included in this endeavor. Simple equations that involve the AC source data, along with the parameters of other important components (particularly the harmonic-generating source), are described. Filter components are determined and tested to meet industry standards' operation performance. Some practical examples are used to illustrate the application of the different filtering schemes.

Because of the expenses incurred in providing harmonic filters, particularly but not exclusively at industrial installations, other methods to alleviate the harmonic distortion problem are often applied. Alternative methods, including use of stiffer AC sources, power converters with increased number of pulses, series reactors, and load reconfiguration, are presented in Chapter 7.

In Chapter 8, a description of the most relevant elements that play a role in the study of the propagation of harmonic currents in a distribution network is presented. These elements include the AC source, transmission lines, cables, transformers, harmonic filters, power factor, capacitor banks, etc. In dealing with the propagation of harmonic currents in electrical networks, it is very important to recognize the complexity that they can reach when extensive networks are considered. Therefore, some examples are illustrated to show the convenience of using specialized tools in the analysis of complicated networks with multiple harmonic sources. The penetration of harmonic currents in the electrical network that can affect adjacent customers and even reach the substation transformer is also discussed.

Finally, a description of the most important aspects to determine power losses in electrical equipment attributed to harmonic waveform distortion is presented in Chapter 9. This is done with particular emphasis on transformers and rotating machines.

Most of the examples presented in this book are based on my experience in industrial applications.

I hope this book provides some useful contribution to the understanding of a complex phenomenon that can assist in the solution of specific problems related to severe waveform distortion in electrical power networks.

Francisco C. De La Rosa

Acknowledgments

My appreciation for the publication of this book goes first to my family for their absolute support. Thanks to Connie, my wife, for bearing with me at all times and especially during the period when this book was written, for the many hours of sleep she lost. Thanks to Eugene, my son, for being patient and considerate with me when I was unable to share much time with him, especially for his positive and thoughtful revision of many parts of the book. His sharp and judicious remarks greatly helped me better describe many of the ideas found in this book.

To produce some of the computer-generated plots presented in the course of the book, I used a number of software tools that were of utmost importance to illustrate fundamental concepts and application examples. Thanks to Professor Mack Grady from the University of Texas at Austin for allowing me to use his HASIP software and to Tom Grebe from Electrotek Concepts, Inc. for granting me permission to use Electrotek Concepts TOP, The Output Processor®. The friendly PSCAD (free) student version from Manitoba HVDC Research Centre Inc. was instrumental in producing many of the illustrations presented in this book and a few examples were also generated with the free Power Quality Teaching Toy Tool from Alex McEachern.

The Author

Francisco De La Rosa, presently a staff scientist at Distribution Control Systems, Inc. (DCSI) in Hazelwood, Missouri, holds BSc and MSc degrees in industrial and power engineering from Coahuila and Monterrey Technological Institutes in Mexico, respectively and a PhD degree in electrical engineering from Uppsala University in Sweden.

Before joining the Advanced Systems and Technology Group at DCSI, an ESCO Technologies Company, Dr. De La Rosa conducted research, tutored, and offered engineering consultancy services for electric, oil, and steel mill companies in the United States, Canada, Mexico, and Venezuela for over 20 years. Dr. De La Rosa taught electrical engineering courses at the Nuevo Leon State University in Monterrey, Mexico as an invited lecturer in 2000–2001. He holds professional membership in the IEEE Power Engineering Society where he participates in working groups dealing with harmonics, power quality, and distributed generation.

Contents

Chapter 1 Fundamentals of Harmonic Distortion and Power Quality
Indices in Electric Power Systems 1

1.1 Introduction 1
1.2 Basics of Harmonic Theory 2
1.3 Linear and Nonlinear Loads 3
    1.3.1 Linear Loads 4
    1.3.2 Nonlinear Loads 6
1.4 Fourier Series 9
    1.4.1 Orthogonal Functions 12
    1.4.2 Fourier Coefficients 13
    1.4.3 Even Functions 13
    1.4.4 Odd Functions 13
    1.4.5 Effect of Waveform Symmetry 14
    1.4.6 Examples of Calculation of Harmonics Using Fourier Series 14
        1.4.6.1 Example 1 14
        1.4.6.2 Example 2 15
1.5 Power Quality Indices under Harmonic Distortion 17
    1.5.1 Total Harmonic Distortion 17
    1.5.2 Total Demand Distortion 17
    1.5.3 Telephone Influence Factor TIF 18
    1.5.4 C Message Index 18
    1.5.5 $I*T$ and $V*T$ Products 18
    1.5.6 K Factor 19
    1.5.7 Displacement, Distortion, and Total Power Factor 19
    1.5.8 Voltage-Related Parameters 20
1.6 Power Quantities under Nonsinusoidal Situations 20
    1.6.1 Instantaneous Voltage and Current 20
    1.6.2 Instantaneous Power 21
    1.6.3 RMS Values 21
    1.6.4 Active Power 21
    1.6.5 Reactive Power 21
    1.6.6 Apparent Power 21
    1.6.7 Voltage in Balanced Three-Phase Systems 22
    1.6.8 Voltage in Unbalanced Three-Phase Systems 23
References 25

Chapter 2 Harmonic Sources 27

2.1 Introduction 27
2.2 The Signature of Harmonic Distortion 28
2.3 Traditional Harmonic Sources 29
    2.3.1 Transformers 36
    2.3.2 Rotating Machines 37
    2.3.3 Power Converters 39
        2.3.3.1 Large Power Converters 45
        2.3.3.2 Medium-Size Power Converters 45
        2.3.3.3 Low-Power Converters 46
        2.3.3.4 Variable Frequency Drives 47
    2.3.4 Fluorescent Lamps 54
    2.3.5 Electric Furnaces 55
2.4 Future Sources of Harmonics 56
References 58

Chapter 3 Standardization of Harmonic Levels 59

3.1 Introduction 59
3.2 Harmonic Distortion Limits 60
    3.2.1 In Agreement with IEEE-519:1992 61
    3.2.2 In Conformance with IEC Harmonic Distortion Limits 63
References 67

Chapter 4 Effects of Harmonics on Distribution Systems 69

4.1 Introduction 69
4.2 Thermal Effects on Transformers 69
    4.2.1 Neutral Conductor Overloading 70
4.3 Miscellaneous Effects on Capacitor Banks 70
    4.3.1 Overstressing 70
    4.3.2 Resonant Conditions 71
    4.3.3 Unexpected Fuse Operation 72
4.4 Abnormal Operation of Electronic Relays 73
4.5 Lighting Devices 73
4.6 Telephone Interference 74
4.7 Thermal Effects on Rotating Machines 74
4.8 Pulsating Torques in Rotating Machines 74
4.9 Abnormal Operation of Solid-State Devices 81
4.10 Considerations for Cables and Equipment Operating in Harmonic Environments 81
    4.10.1 Generators 81
    4.10.2 Conductors 83
    4.10.3 Energy-Metering Equipment 83
References 83

Chapter 5 Harmonics Measurements 85

5.1 Introduction 85
5.2 Relevant Harmonic Measurement Questions 86
    5.2.1 Why Measure Waveform Distortion 86
    5.2.2 How to Carry out Measurements 87
    5.2.3 What Is Important to Measure 87
    5.2.4 Where Should Harmonic Measurements Be Conducted 88
    5.2.5 How Long Should Measurements Last 88
5.3 Measurement Procedure 89
    5.3.1 Equipment 89
    5.3.2 Transducers 90
5.4 Relevant Aspects 90
References 91

Chapter 6 Harmonic Filtering Techniques 93

6.1 Introduction 93
6.2 General Aspects in the Design of Passive Harmonic Filters 93
6.3 Single-Tuned Filters 94
    6.3.1 Design Equations for the Single-Tuned Filter 96
    6.3.2 Parallel Resonant Points 97
    6.3.3 Quality Factor 100
    6.3.4 Recommended Operation Values for Filter Components 101
        6.3.4.1 Capacitors 101
        6.3.4.2 Tuning Reactor 104
    6.3.5 Unbalance Detection 104
    6.3.6 Filter Selection and Performance Assessment 104
6.4 Band-Pass Filters 105
6.5 Relevant Aspects to Consider in the Design of Passive Filters 107
6.6 Methodology for Design of Tuned Harmonic Filters 108
    6.6.1 Select Capacitor Bank Needed to Improve the Power Factor from the Present Level Typically to around 0.9 to 0.95 108
    6.6.2 Choose Reactor that, in Series with Capacitor, Tunes Filter to Desired Harmonic Frequency 109
    6.6.3 Determine Whether Capacitor-Operating Parameters Fall within IEEE-18[2] Maximum Recommended Limits 109
        6.6.3.1 Capacitor Voltage 109
        6.6.3.2 Current through the Capacitor Bank 110
        6.6.3.3 Determine the Capacitor Bank Duty and Verify that It Is within Recommended IEEE-18 Limits 110
    6.6.4 Test Out Resonant Conditions 110
6.7 Example 1: Adaptation of a Power Factor Capacitor Bank into a Fifth Harmonic Filter 110
6.8 Example 2: Digital Simulation of Single-Tuned Harmonic Filters 113

6.9   Example 3: High-Pass Filter at Generator Terminals Used to Control a Resonant Condition 117
6.10 Example 4: Comparison between Several Harmonic Mitigating Schemes Using University of Texas at Austin HASIP Program 124
References 129

Chapter 7   Other Methods to Decrease Harmonic Distortion Limits 131

7.1   Introduction 131
7.2   Network Topology Reconfiguration 132
7.3   Increase of Supply Mode Stiffness 132
7.4   Harmonic Cancellation through Use of Multipulse Converters 134
7.5   Series Reactors as Harmonic Attenuator Elements 135
7.6   Phase Balancing 136
      7.6.1   Phase Voltage Unbalance 137
      7.6.2   Effects of Unbalanced Phase Voltage 137
Reference 138

Chapter 8   Harmonic Analyses 139

8.1   Introduction 139
8.2   Power Frequency vs. Harmonic Current Propagation 139
8.3   Harmonic Source Representation 142
      8.3.1   Time/Frequency Characteristic of the Disturbance 142
      8.3.2   Resonant Conditions 147
      8.3.3   Burst-Type Harmonic Representation 148
8.4   Harmonic Propagation Facts 149
8.5   Flux of Harmonic Currents 150
      8.5.1   Modeling Philosophy 151
      8.5.2   Single-Phase vs. Three-Phase Modeling 152
      8.5.3   Line and Cable Models 152
      8.5.4   Transformer Model for Harmonic Analysis 153
      8.5.5   Power Factor Correction Capacitors 154
8.6   Interrelation between AC System and Load Parameters 154
      8.6.1   Particulars of Distribution Systems 156
      8.6.2   Some Specifics of Industrial Installations 157
8.7   Analysis Methods 158
      8.7.1   Simplified Calculations 158
      8.7.2   Simulation with Commercial Software 159
8.8   Examples of Harmonic Analysis 160
      8.8.1   Harmonic Current during Transformer Energization 160
      8.8.2   Phase A to Ground Fault 160
References 167

Chapter 9 Fundamentals of Power Losses in Harmonic Environments 169

9.1 Introduction 169
9.2 Meaning of Harmonic-Related Losses 169
9.3 Relevant Aspects of Losses in Power Apparatus and Distribution Systems 171
9.4 Harmonic Losses in Equipment 172
    9.4.1 Resistive Elements 172
    9.4.2 Transformers 174
        9.4.2.1 Crest Factor 174
        9.4.2.2 Harmonic Factor or Percent of Total Harmonic Distortion 175
        9.4.2.3 K Factor 175
9.5 Example of Determination of K Factor 176
9.6 Rotating Machines 177
References 179

Index 181

1 Fundamentals of Harmonic Distortion and Power Quality Indices in Electric Power Systems

1.1 INTRODUCTION

Ideally, an electricity supply should invariably show a perfectly sinusoidal voltage signal at every customer location. However, for a number of reasons, utilities often find it hard to preserve such desirable conditions. The deviation of the voltage and current waveforms from sinusoidal is described in terms of the waveform distortion, often expressed as harmonic distortion.

Harmonic distortion is not new and it constitutes at present one of the main concerns for engineers in the several stages of energy utilization within the power industry. In the first electric power systems, harmonic distortion was mainly caused by saturation of transformers, industrial arc furnaces, and other arc devices like large electric welders. The major concern was the effect that harmonic distortion could have on electric machines, telephone interference, and increased risk of faults from overvoltage conditions developed on power factor correction capacitors

In the past, harmonics represented less of a problem due to the conservative design of power equipment and to the common use of delta-grounded wye connections in distribution transformers.

The increasing use of nonlinear loads in industry is keeping harmonic distortion in distribution networks on the rise. The most used nonlinear device is perhaps the static power converter so widely used in industrial applications in the steel, paper, and textile industries. Other applications include multipurpose motor speed control, electrical transportation systems, and electrodomestic appliances. By 2000, it was estimated that electronic loads accounted for around half of U.S. electrical demand, and much of that growth in electronic load involved the residential sector.[1]

A situation that has raised waveform distortion levels in distribution networks even further is the application of capacitor banks used in industrial plants for power factor correction and by power utilities for increasing voltage profile along distribution lines. The resulting reactive impedance forms a tank circuit with the system inductive reactance at a certain frequency likely to coincide with one of the characteristic harmonics of the load. This condition will trigger large oscillatory currents and voltages that may stress the insulation. This situation imposes a serious challenge to industry and utility engineers to pinpoint and to correct excessive harmonic waveform distortion levels on the waveforms because its steady increase happens to take place right at the time when the use of sensitive electronic equipment is on the rise.

No doubt harmonic studies from the planning to the design stages of power utility and industrial installations will prove to be an effective way to keep networks and equipment under acceptable operating conditions and to anticipate potential problems with the installation or addition of nonlinear loads.

1.2 BASICS OF HARMONIC THEORY

The term "harmonics" was originated in the field of acoustics, where it was related to the vibration of a string or an air column at a frequency that is a multiple of the base frequency. A harmonic component in an AC power system is defined as a sinusoidal component of a periodic waveform that has a frequency equal to an integer multiple of the fundamental frequency of the system.

Harmonics in voltage or current waveforms can then be conceived as perfectly sinusoidal components of frequencies multiple of the fundamental frequency:

$$f_h = (h) \times (\text{fundamental frequency}) \tag{1.1}$$

where h is an integer.

For example, a fifth harmonic would yield a harmonic component:

$$f_h = (5) \times (60) = 300 \text{ Hz and } f_h = (5) \times (50) = 250 \text{ Hz}$$

in 60- and 50-Hz systems, respectively.

Figure 1.1 shows an ideal 60-Hz waveform with a peak value of around 100 A, which can be taken as one per unit. Likewise, it also portrays waveforms of amplitudes (1/7), (1/5), and (1/3) per unit and frequencies seven, five, and three times the fundamental frequency, respectively. This behavior showing harmonic components of decreasing amplitude often following an inverse law with harmonic order is typical in power systems.

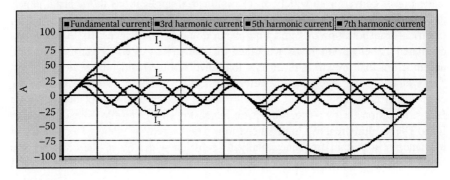

FIGURE 1.1 Sinusoidal 60-Hz waveform and some harmonics.

FIGURE 1.2 Sinusoidal waveform distorted by third, fifth, and seventh harmonics.

These waveforms can be expressed as:

$$i_1 = Im_1 \sin\omega t \tag{1.2}$$

$$i_3 = Im_3 \sin(3\omega t - \delta_3) \tag{1.3}$$

$$i_5 = Im_5 \sin(5\omega t - \delta_5) \tag{1.4}$$

$$i_7 = Im_7 \sin(7\omega t - \delta_7) \tag{1.5}$$

where Im_h is the peak RMS value of the harmonic current h.

Figure 1.2 shows the same harmonic waveforms as those in Figure 1.1 superimposed on the fundamental frequency current yielding I_{total}. If we take only the first three harmonic components, the figure shows how a distorted current waveform at the terminals of a six-pulse converter would look. There would be additional harmonics that would impose a further distortion.

The resultant distorted waveform can thus be expressed as:

$$I_{total} = Im_1 \sin\omega t + Im_3 \sin(3\omega t - \delta_3) + Im_5 \sin(5\omega t - \delta_5) + Im_7 \sin(7\omega t - \delta_7) \tag{1.6}$$

In this way, a summation of perfectly sinusoidal waveforms can give rise to a distorted waveform. Conversely, a distorted waveform can be represented as the superposition of a fundamental frequency waveform with other waveforms of different harmonic frequencies and amplitudes.

1.3 LINEAR AND NONLINEAR LOADS

From the discussion in this section, it will be evident that a load that draws current from a sinusoidal AC source presenting a waveform like that of Figure 1.2 cannot be conceived as a linear load.

Resistive elements	Inductive elements	Capacitive elements
• Incandescent lighting • Electric heaters	• Induction motors • Current limiting reactors • Induction generators (wind mills) • Damping reactors used to attenuate harmonics • Tuning reactors in harmonic filters	• Power factor correction capacitor banks • Underground cables • Insulated cables • Capacitors used in harmonic filters

FIGURE 1.3 Examples of linear loads.

1.3.1 Linear Loads

Linear loads are those in which voltage and current signals follow one another very closely, such as the voltage drop that develops across a constant resistance, which varies as a direct function of the current that passes through it. This relation is better known as Ohm's law and states that the current through a resistance fed by a varying voltage source is equal to the relation between the voltage and the resistance, as described by:

$$i(t) = \frac{v(t)}{R} \qquad (1.7)$$

This is why the voltage and current waveforms in electrical circuits with linear loads look alike. Therefore, if the source is a clean open circuit voltage, the current waveform will look identical, showing no distortion. Circuits with linear loads thus make it simple to calculate voltage and current waveforms. Even the amounts of heat created by resistive linear loads like heating elements or incandescent lamps can easily be determined because they are proportional to the square of the current. Alternatively, the involved power can also be determined as the product of the two quantities, voltage and current.

Other linear loads, such as electrical motors driving fans, water pumps, oil pumps, cranes, elevators, etc., not supplied through power conversion devices like variable frequency drives or any other form or rectification/inversion of current will incorporate magnetic core losses that depend on iron and copper physical characteristics. Voltage and current distortion may be produced if ferromagnetic core equipment is operated on the saturation region, a condition that can be reached, for instance, when equipment is operated above rated values.

Capacitor banks used for power factor correction by electric companies and industry are another type of linear load. Figure 1.3 describes a list of linear loads.

A voltage and current waveform in a circuit with linear loads will show the two waveforms in phase with one another. Voltage and current involving inductors make voltage lead current and circuits that contain power factor capacitors make current lead voltage. Therefore, in both cases, the two waveforms will be out of phase from one another. However, no waveform distortion will take place.

Fundamentals of Harmonic Distortion and Power Quality Indices

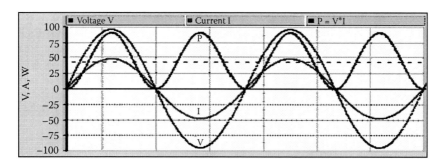

FIGURE 1.4 Relation among voltage, current, and power in a purely resistive circuit.

Figure 1.4 presents the relation among voltage, current, and power in a linear circuit consisting of an AC source feeding a purely resistive circuit. Notice that instantaneous power, $P = V * I$, is never negative because both waveforms are in phase and their product will always yield a positive quantity. The same result is obtained when power is obtained as the product of the resistance with the square of the current.

Figure 1.5(a) shows the relation between the same parameters for the case when current I lags the voltage V, which would correspond to an inductive load, and Figure 1.5(b) for the case when I leads the voltage V as in the case of a capacitive load.

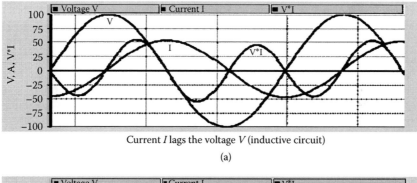

Current I lags the voltage V (inductive circuit)

(a)

Current I leads the voltage V (capacitive circuit)

(b)

FIGURE 1.5 Relation among voltage, current, and their product in inductive (a) and capacitive (b) circuits, respectively.

Negative and positive displacement power factors (discussed in Section 1.5) are related to Figure 1.5(a) and 1.5(b), respectively. Note that in these cases the product $V * I$ has positive and negative values. The positive values correspond to the absorption of current by the load and the negative values to the flux of current towards the source.

In any case, the sinusoidal nature of voltage and current waveforms is preserved, just as in the case of Figure 1.4 that involves a purely resistive load. Observe that even the product $V * I$ has equal positive and negative cycles with a zero average value; it is positive when V and I are positive and negative when V or I are negative.

1.3.2 NONLINEAR LOADS

Nonlinear loads are loads in which the current waveform does not resemble the applied voltage waveform due to a number of reasons, for example, the use of electronic switches that conduct load current only during a fraction of the power frequency period. Therefore, we can conceive nonlinear loads as those in which Ohm's law cannot describe the relation between V and I. Among the most common nonlinear loads in power systems are all types of rectifying devices like those found in power converters, power sources, uninterruptible power supply (UPS) units, and arc devices like electric furnaces and fluorescent lamps. Figure 1.6 provides a more extensive list of various devices in this category. As later discussed in Chapter 4, **nonlinear loads** cause a number of disturbances like voltage waveform distortion, overheating in transformers and other power devices, overcurrent on equipment-neutral connection leads, telephone interference, and microprocessor control problems, among others.

Figure 1.7 shows the voltage and current waveforms during the switching action of an insulated gate bipolar transistor (IGBT), a common power electronics solid-state device. This is the simplest way to illustrate the performance of a nonlinear load in which the current does not follow the sinusoidal source voltage waveform except during the time when firing pulses FP1 and FT2 (as shown on the lower plot) are ON. Some motor speed controllers, household equipment like TV sets and VCRs,

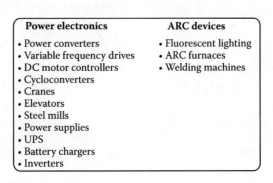

FIGURE 1.6 Examples of some nonlinear loads.

FIGURE 1.7 Relation between voltage and current in a typical nonlinear power source.

and a large variety of other residential and commercial electronic equipment use this type of voltage control. When the same process takes place in three-phase equipment and the amount of load is significant, a corresponding distortion can take place also in the voltage signal.

Even linear loads like power transformers can act nonlinear under saturation conditions. What this means is that, in certain instances, the magnetic flux density (B) in the transformer ceases to increase or increases very little as the magnetic flux intensity (H) keeps growing. This occurs beyond the so-called saturation knee of the magnetizing curve of the transformer. The behavior of the transformer under changing cycles of positive and negative values of H is shown in Figure 1.8 and is known as hysteresis curve.

Of course, this nonlinear effect will last as long as the saturation condition prevails. For example, an elevated voltage can be fed to the transformer during

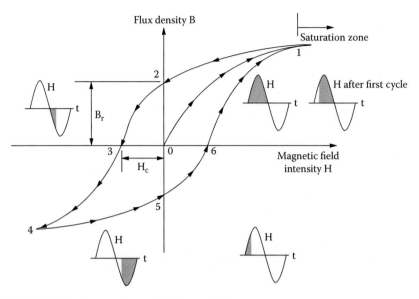

FIGURE 1.8 Transformer hysteresis characteristic.

low-load conditions that can last up to several hours, but an overloaded transformer condition is often observed during starting of large motors or high inertia loads in industrial environments lasting a few seconds. The same situation can occur practically with other types of magnetic core devices.

In Figure 1.8, the so-called transformer magnetizing curve of the transformer (curve 0–1) starts at point 0 with the increase of the magnetic field intensity H, reaching point 1 at peak H, beyond which the magnetic flux shows a flat behavior, i.e., a small increase in B on a large increase in H. Consequently, the current starts getting distorted and thus showing harmonic components on the voltage waveform too. Notice that from point 1 to point 2, the B–H characteristic follows a different path so that when magnetic field intensity has decreased to zero, a remanent flux density, B_r, called *permanent magnetization* or *remanence* is left in the transformer core. This is only cancelled when electric field intensity reverses and reaches the so-called *coercive force* H_c. Point 4 corresponds to the negative cycle magnetic field intensity peak. When H returns to zero at the end of the first cycle, the B–H characteristic ends in point 5. From here a complete hysteresis cycle would be completed when H reaches again its peak positive value to return to point 1.

The area encompassed by the hysteresis curve is proportional to the transformer core losses. It is important to note that transformer cores that offer a small coercive force would be needed to minimize losses.

Note that the normal operation of power transformers should be below the saturation region. However, when the transformer is operated beyond its rated power (during peak demand hours) or above nominal voltage (especially if power factor capacitor banks are left connected to the line under light load conditions), transformers are prone to operate under saturation.

Fundamentals of Harmonic Distortion and Power Quality Indices

Practically speaking, all transformers reach the saturation region on energization, developing large inrush (magnetizing) currents. Nevertheless, this is a condition that lasts only a few cycles. Another situation in which the power transformer may operate on the saturation region is under unbalanced load conditions; one of the phases carries a different current than the other phases, or the three phases carry unlike currents.

1.4 FOURIER SERIES

By definition, a periodic function, $f(t)$, is that where $f(t) = f(t + T)$. This function can be represented by a trigonometric series of elements consisting of a DC component and other elements with frequencies comprising the fundamental component and its integer multiple frequencies. This applies if the following so-called Dirichlet conditions[2] are met:

If a discontinuous function, $f(t)$ has a finite number of discontinuities over the period T
If $f(t)$ has a finite mean value over the period T
If $f(t)$ has a finite number of positive and negative maximum values

The expression for the trigonometric series $f(t)$ is as follows:

$$f(t) = \frac{a_0}{2} + \sum_{h=1}^{\infty}\left[a_h \cos(h\omega_0 t) + b_h \sin(h\omega_0 t)\right] \quad (1.8)$$

where $\omega_0 = 2\pi/T$.

We can further simplify Equation (1.8), which yields:

$$f(t) = c_0 + \sum_{h=1}^{\infty} c_h \sin(h\omega_0 t + \phi_h) \quad (1.9)$$

where

$$c_0 = \frac{a_0}{2}, c_h = \sqrt{a_h^2 + b_h^2}, and \phi_h = \tan_{-1}\left(\frac{a_h}{b_h}\right)$$

Equation (1.9) is known as a Fourier series and it describes a periodic function made up of the contribution of sinusoidal functions of different frequencies.

- $(h\ \omega_0)$ hth order harmonic of the periodic function
- c_0 magnitude of the DC component
- c_h and ϕ_h magnitude and phase angle of the hth harmonic component

The component with $h = 1$ is called the fundamental component. Magnitude and phase angle of each harmonic determine the resultant waveform $f(t)$.

Equation (1.8) can be represented in a complex form as:

$$f(t) = \sum_{h=1}^{\infty} c_h e^{jh\omega_0 t} \qquad (1.10)$$

where $h = 0, \pm 1, \pm 2, \ldots$

$$c_h = \frac{1}{T} \int_{-T/2}^{T/2} f(t) e^{-jh\omega_0 t} \, dt \qquad (1.11)$$

Generally, the frequencies of interest for harmonic analysis include up to the 40th or so harmonics.[3]

The main source of harmonics in power systems is the static power converter. Under ideal operation conditions, harmonics generated by a p pulse power converter are characterized by:

$$Ih = \frac{I_1}{h}, \text{ and } h = pn \pm 1 \qquad (1.12)$$

where h stands for the characteristic harmonics of the load; $n = 1, 2, \ldots$; and p is an integer multiple of six.

A bar plot of the amplitudes of harmonics generated in a six-pulse converter normalized as c_n/c_1 is called the harmonic spectrum, and it is shown in Figure 1.9.

FIGURE 1.9 Example of a harmonic spectrum.

FIGURE 1.10 Decomposition of a distorted waveform.

The breakdown of the current waveform including the four dominant harmonics is shown in Figure 1.10. Notice that the harmonic spectrum is calculated with the convenient Electrotek Concepts TOP Output Processor.[4]

Noncharacteristic harmonics appear when:

The input voltages are unbalanced.
The commutation reactance between phases is not equal.
The "space" between triggering pulses at the converter rectifier is not equal.

These harmonics are added together with the characteristic components and can produce waveforms with components that are not integer multiples of the fundamental frequency in the power system, also known as interharmonics.

A main source of interharmonics is the AC to AC converter, also called cycloconverter. These devices have a fixed amplitude and frequency at the input; at the output, amplitude and frequency can be variable. A typical application of a cycloconverter is as an AC traction motor speed control and other high-power, low-frequency applications, generally in the MW range.

Other important types of harmonics are those produced by electric furnaces, usually of a frequency lower than that of the AC system. These are known as subharmonics and are responsible for the light flickering phenomenon visually perceptible in incandescent and arc-type lighting devices.

Odd multiples of three (triplen) harmonics in balanced systems can be blocked using ungrounded neutral or delta-connected transformers because these are zero sequence harmonics. This is why triplen harmonics are often ignored in harmonic studies.

1.4.1 Orthogonal Functions

A set of functions, ϕ_i, defined in $a \leq x \leq b$ is called orthogonal (or unitary, if complex) if it satisfies the following condition:

$$\int_a^b \phi_i(x)\phi_j^*(x)dx = K_i \delta_{ij} \tag{1.13}$$

where $\delta_{ij} = 1$ for $i = j$, and $= 0$ for $i \neq j$, and $*$ is the complex conjugate.

It can also be shown that the functions:

$$\{1, \cos(\omega_0 t), \ldots, \sin(\omega_0 t), \ldots, \cos(h\omega_0 t), \ldots, \sin(h\omega_0 t), \ldots\} \tag{1.14}$$

for which the following conditions are valid:

$$\int_{-T/2}^{T/2} \cos kx \cos lx \, dx = \begin{cases} 0, k \neq l, \\ \pi, k = l, \end{cases} \tag{1.15}$$

$$\int_{-T/2}^{T/2} \sin kx \sin lx \, dx = \begin{cases} 0, k \neq l, \\ \pi, k = l, \end{cases} \tag{1.16}$$

$$\int_{-T/2}^{T/2} \cos kx \sin lx \, dx = 0 \quad (k = 1, 2, 3, \ldots), \tag{1.17}$$

$$\int_{-T/2}^{T/2} \cos kx \, dx = 0 \quad (k = 1, 2, 3, \ldots), \tag{1.18}$$

$$\int_{-T/2}^{T/2} \sin kx \, dx = 0 \quad (k = 1, 2, 3, \ldots), \tag{1.19}$$

Fundamentals of Harmonic Distortion and Power Quality Indices

FIGURE 1.11 Example of even functions.

$$\int_{-T/2}^{T/2} 1 dx = 2\pi \tag{1.20}$$

are a set of orthogonal functions. From Equation (1.14) to Equation (1.20), it is clear that the integral over the period ($-\pi$ to π) of the product of any two sine and cosine functions is zero.

1.4.2 FOURIER COEFFICIENTS

Integrating Equation (1.8) and applying the orthogonal functions (Equation 1.15 through Equation 1.20), we obtain the Fourier coefficients as follows:

$$a_0 = \frac{2}{T} \int_{-T/2}^{T/2} f(t) dt, \tag{1.21}$$

$$a_h = \frac{2}{T} \int_{-T/2}^{T/2} f(t) \cos(h\omega_0 t) dt, \quad \text{and,} \tag{1.22}$$

$$b_h = \frac{2}{T} \int_{-T/2}^{T/2} f(t) \sin(h\omega_0 t) dt \tag{1.23}$$

where $h = 1, 2, \ldots \infty$.

1.4.3 EVEN FUNCTIONS

A function, $f(t)$, is called an even function if it has the following property:

$$f(t) = f(-t) \tag{1.24}$$

Figure 1.11 shows some examples of even functions.

1.4.4 ODD FUNCTIONS

A function is called an odd function if:

FIGURE 1.12 Example of odd functions.

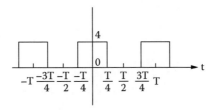

FIGURE 1.13 Square wave function.

$$f(-t) = -f(t) \quad (1.25)$$

as portrayed in Figure 1.12.

An even function is symmetrical to the vertical axis at the origin and an odd function is asymmetrical to the vertical axis at the origin. A function with a period, T, is half-wave symmetrical if it satisfies the condition:

$$f(t) = -f[t \pm (T/2)] \quad (1.26)$$

1.4.5 Effect of Waveform Symmetry

The Fourier series of an even function contain only cosine terms and may also include a DC component. Thus, the coefficients b_i are zero.

The Fourier series of an odd function contain only sine terms. The coefficients a_i are all zero.

The Fourier series of a function with half–wave symmetry contain only odd harmonic terms with $a_i = 0$ for $i = 0$ and all other even terms and $b_i = 0$ for all even values of i.

1.4.6 Examples of Calculation of Harmonics Using Fourier Series

1.4.6.1 Example 1

Consider the periodic function of Figure 1.13, which can be expressed as follows:

$$0, \; -T/2 < t < -T/4 \quad (1.27)$$

Fundamentals of Harmonic Distortion and Power Quality Indices

$$f(t) = 4, -T/4 < t < T/4 \quad (1.28)$$

$$0, T/4 < t < T/2 \quad (1.29)$$

for which we can calculate the Fourier coefficients using Equation (1.21) through Equation (1.23) as follows:

$$a_0 = \frac{2}{T}\left(\int_{-T/2}^{T/2} f(t)dt\right) = \frac{2}{T}\left(\int_{-T/2}^{-T/4} 0 \cdot dt + \int_{-T/4}^{T/4} 4 \cdot dt + \int_{T/4}^{T/2} 0 \cdot dt\right) \quad (1.30)$$

$$= \frac{2}{T}\left[4(T/4+T/4)\right] = 4$$

$$a_1 = \frac{2}{T}\int_{-T/2}^{T/2} f(t)\cos(w_0 t)dt =$$

$$= \frac{2}{T}\left(\int_{-T/2}^{-T/4} 0 \cdot \cos(\omega_0 t)dt + \int_{-T/4}^{T/4} 4 \cdot \cos(\omega_0 t)dt + \int_{T/4}^{T/2} 0 \cdot \cos(\omega_0 t)dt\right) = \quad (1.31)$$

$$= \frac{8}{\omega_0 T}(\sin \omega_0 t)\Big|_{-T/4}^{T/4} = \frac{8}{2\pi}\left[\sin\left(\frac{\pi}{2}\right) + \sin\left(\frac{\pi}{2}\right)\right] = \frac{8}{\pi}$$

We equally find that:

$$a_i = \begin{cases} 0 \text{--------------} i=even \\ -1^{(i-1)/2}\frac{8}{i\pi} \text{------} i=odd \end{cases} \quad b_i = 0 \quad (1.32)$$

Therefore, from Equation (1.8), the Fourier series of this waveform is as follows:

$$f(t) = 2 + \frac{8}{\pi}\left(\cos \pi t - \frac{1}{3}\cos 3\pi t + \frac{1}{5}\cos 5\pi t - \ldots\right) \quad (1.33)$$

1.4.6.2 Example 2

Consider now that the periodic function of Figure 1.13 has its origin shifted one fourth of a cycle, as illustrated in Figure 1.14.

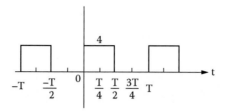

FIGURE 1.14 Square wave function shifted one fourth of a cycle relative to Figure 1.13.

$$a_0 = \frac{2}{T}\left(\int_{-T/2}^{T/2} f(t)dt\right) =$$

(1.34)

$$= \frac{2}{T}\left(\int_{-T/2}^{0} 0 \cdot dt + \int_{0}^{T/2} 4 \cdot dt\right) = \frac{2}{T}\left[4(T/2-0)\right] = 4$$

$$a_1 = \frac{2}{T}\int_{-T/2}^{T/2} f(t)\cos(w_0 t)dt =$$

$$= \frac{2}{T}\left(\int_{-T/2}^{-0} 0 \cdot \cos(\omega_0 t)dt + \int_{0}^{T/2} 4 \cdot \cos(\omega_0 t)dt\right) =$$

(1.35)

$$\frac{8}{\omega_0 T}(\sin \omega_0 t)\Big|_{0}^{T/2} = \frac{8}{2\pi}\left[\sin\left(\frac{2\pi}{2}\right) - \sin(0)\right] = 0$$

Applying the orthogonality relations to Equation (1.22), we find that all a_i coefficients are zero. If we now try Equation (1.23), we determine the coefficients associated with the sine function in this series. For example, the first term, b_1, is calculated as follows:

$$b_1 = \frac{2}{T}\int_{0}^{T} f(t)\sin(\omega_0)t dt = \frac{2\omega_0}{2\pi}\int_{0}^{T/2} 4\sin(\omega_0 t)dt = -\left(\frac{2}{T}\right)\frac{4\cos(\omega_0 t)}{\omega_0}\Big|_{0}^{T/2} = \frac{8}{\pi}$$

(1.36)

Likewise, we find that:

Fundamentals of Harmonic Distortion and Power Quality Indices

$$b_i = \begin{cases} 0 \text{------} i=even \\ \dfrac{4}{i\pi} \text{------} i=odd \end{cases} \quad (1.37)$$

Therefore, following Equation (1.8), the Fourier series of this waveform reduced to its first three terms is as follows:

$$f(t) = 2 + \frac{8}{\pi}\left(\sin \pi t - \frac{1}{3}\sin 3\pi t + \frac{1}{5}\sin 5\pi t - \ldots\right) \quad (1.38)$$

1.5 POWER QUALITY INDICES UNDER HARMONIC DISTORTION

This section describes power quality indices that are comprehensibly described in reference 5 and reference 6.

1.5.1 TOTAL HARMONIC DISTORTION

Total harmonic distortion (THD) is an important index widely used to describe power quality issues in transmission and distribution systems. It considers the contribution of every individual harmonic component on the signal. THD is defined for voltage and current signals, respectively, as follows:

$$THD_V = \frac{\sqrt{\sum_{h=2}^{\infty} V_h^2}}{V_1} \quad (1.39)$$

$$THD_I = \frac{\sqrt{\sum_{h=2}^{\infty} I_h^2}}{I_1} \quad (1.40)$$

This means that the ratio between rms values of signals including harmonics and signals considering only the fundamental frequency define the total harmonic distortion.

1.5.2 TOTAL DEMAND DISTORTION

Harmonic distortion is most meaningful when monitored at the point of common coupling (PCC) — usually the customer's metering point — over a period that can reflect maximum customer demand, typically 15 to 30 minutes as suggested in Standard IEEE-519.[7] Weak sources with a large demand current relative to their rated current will tend to show greater waveform distortion. Conversely, stiff sources characterized for operating at low demand currents will show decreased waveform distortion. The total demand distortion is based on the demand current, I_L, over the monitoring period:

$$TDD = \frac{\sqrt{\sum_{h=2}^{\infty} I_h^2}}{I_L} \quad (1.41)$$

1.5.3 TELEPHONE INFLUENCE FACTOR TIF

This index is found in IEEE-519[7] as a measure of audio circuit interference produced by harmonics in electric power systems. It will thus use the total harmonic distortion concept influenced by appropriate weighting factors, ω_h, that establish the sensitivity of the human ear to noise from different frequencies:

$$TIF = \frac{\sqrt{\sum_{h=2}^{\infty} w_h^2 I_h^2}}{I_{rms}} \quad (1.42)$$

1.5.4 C MESSAGE INDEX

This index is related in reference 7 to communication interference:

$$C_I = \frac{\sqrt{\sum_{h=2}^{\infty} c_h^2 I_h}}{I_{rms}} \quad (1.43)$$

where c_h is the weighting factor, ω_h, divided by five times the harmonic order h.

1.5.5 $I * T$ AND $V * T$ PRODUCTS

These indices are used as another measure of harmonic interference in audio circuits. Because of their intimate relation with total waveform distortion, $I * T$ and $V * T$ are also indicative of shunt capacitor stress and voltage distortion, respectively:

$$I * T = TIF * I_{rms} = \sqrt{\sum_{h=2}^{\infty} (w_h I_h)^2} \quad (1.44)$$

$$V * T = TIF * V_{rms} = \sqrt{\sum_{h=2}^{\infty} (w_h V_h)^2} \quad (1.45)$$

1.5.6 K Factor

This is a useful index intended to follow the requirements of the National Electrical Code (NEC) and Underwriter's Laboratories (UL), (well summarized by its originator, Frank[8]) regarding the capability of distribution and special application transformers in industry to operate within specified thermal limits in harmonic environments. These are transformers designed to operate at lower flux densities than conventional designs to allow for the additional flux produced by (largely the third) harmonic currents. Also, to reduce the Eddy or circulating current losses in the core, strip windings, interleaving windings, and transposition conductors are used. The formula used to calculate the K factor (as presented in the IEEE Tutorial Modeling and Simulations[5]) is as follows:

$$K = \frac{\sum_{h=1}^{\infty} h^2 \left(\frac{I_h}{I_1}\right)^2}{\sum_{h=1}^{\infty} \left(\frac{I_h}{I_1}\right)^2} = \qquad (1.46)$$

$$= \sum_{h=1}^{\infty} \left[I_h(p.u.)\right]^2 (h^2)$$

1.5.7 Displacement, Distortion, and Total Power Factor

With an increasing harmonic distortion environment, the conventional definition of power factor as the cosine of the angle between fundamental frequency voltage and current has progressed to consider the signal's rms values, which make up the contribution of components of different frequencies. Thus, displacement power factor (DPF) continues to characterize the power frequency factor, while distortion (or true) power factor (TPF) emerges as the index that tracks rms signal variations. Total power factor (PFtotal) thus becomes the product of distortion and true power factors:

$$PFtotal = DPF * TPF = \cos(\theta 1 - \delta 1) * \frac{P_{total}}{S_{total}} =$$

$$= \left(\frac{P_1}{V_1 I_1}\right) \left(\frac{\sum_{h=1}^{\infty} V_h I_h \cos(\theta_h - \delta_h)}{\sqrt{\sum_{h=1}^{\infty} (V_h)^2} \sqrt{\sum_{h=1}^{\infty} (I_h)^2}}\right) = \qquad (1.47)$$

$$= \left(\frac{P_1}{V_1 I_1}\right) \sqrt{\frac{1}{1 + \left(\frac{THD_I}{100}\right)^2}}$$

where P_1, V_1, and I_1 are fundamental frequency quantities and V_h, I_h, θ_h, and δ_h are related to a frequency, h, times the system power frequency.

Because true power factor is always less than unity, it also holds that:

$$PF_{total} \leq DPF \qquad (1.48)$$

In Equation (1.47), note that fundamental displacement power factor is the ratio between P_{total}/S_{total} or $P_1/(V_1 I_1)$.

1.5.8 VOLTAGE-RELATED PARAMETERS

Crest factor, unbalance factor and flicker factor are intended for assessing dielectric stress, three-phase circuit balance, and source stiffness with regard to its capability of maintaining an adequate voltage regulation, respectively:

$$CrestFactor = \frac{Vpeak}{Vrms} \qquad (1.49)$$

$$UnbalanceFactor = \frac{|V_-|}{|V_+|} \qquad (1.50)$$

$$FlickerFactor = \frac{\Delta V}{|V|} \qquad (1.51)$$

1.6 POWER QUANTITIES UNDER NONSINUSOIDAL SITUATIONS

This section describes IEEE quantities under nonsinusoidal situations.[5] It is noteworthy to emphasize that all quantities referred to in this section are based on the trigonometric Fourier series definition described in Section 1.4 as Equation (1.9). These quantities are expressed in a way that they account for the contribution of individual harmonic frequency components. In this section, $f(t)$ represents instantaneous voltage or current as a function of time; F_h is the peak value of the signal component of harmonic frequency h.

1.6.1 INSTANTANEOUS VOLTAGE AND CURRENT

$$f(t) = \sum_{h=1}^{\infty} f_h(t) = \sum_{h=1}^{\infty} \sqrt{2} F_h \sin(h\omega_0 t + \theta_h) \qquad (1.52)$$

Fundamentals of Harmonic Distortion and Power Quality Indices

1.6.2 Instantaneous Power

$$p(t) = v(t)i(t) \tag{1.53}$$

1.6.3 RMS Values

$$F_{rms} = \sqrt{\frac{1}{T}\int_0^T f^2(t)dt} = \sqrt{\sum_{h=1}^{\infty} F_h^2} \tag{1.54}$$

where F_{rms} is the root mean square of function F, which in our case can be voltage or current.

1.6.4 Active Power

Every harmonic provides a contribution to the average power that can be positive or negative. However, the resultant harmonic power is very small relative to the fundamental frequency active power.

$$P = \frac{1}{T}\int_0^T p(t)dt = \sum_{h=1}^{\infty} V_h I_h \cos(\theta_h - \delta_h) = \sum_{h=1}^{\infty} P_h \tag{1.55}$$

1.6.5 Reactive Power

$$Q = \frac{1}{T}\int_0^T q(t)dt = \sum_{h=1}^{\infty} V_h I_h \sin(\theta_h - \delta_h) = \sum_{h=1}^{\infty} Q_h \tag{1.56}$$

1.6.6 Apparent Power

Many arguments have been written about the most suitable way to express the apparent power under the presence of harmonic distortion. A good summary of such efforts can be found in Arrillaga,[6] who refers to the initial approach by Budeanu,[9] Fryze,[10] and Kusters and Moore[11] and the most recent work by Emanuel.[12,13] Arrillaga also shows how all formulations lead to somewhat different results in the determination of reactive power quantities under waveform distortion. An expression generally accepted by IEEE and IEC is that proposed by Budeanu in Antoniu[9]:

$$S^2 = P^2 + \sum_{i=1}^{n} V_1 I_1 \sin(\varphi_1) + D^2 \tag{1.57}$$

For three-phase systems, the per-phase (k) vector apparent power, S_v, as proposed in Frank,[8] can be expressed, as adapted from Arrillaga,[7] as follows:

$$S_v = \sqrt{\left(\sum_k P_k\right)^2 + \left(\sum_k Q_{bk}\right)^2 + \left(\sum_k D_k\right)^2} \qquad (1.58)$$

and the arithmetic apparent power, S_a, as:

$$S_a = \sum_k \sqrt{P_k^2 + Q_{bk}^2 + D_k^2} \qquad (1.59)$$

where P, Q_b, and D are the active, reactive, and distortion orthogonal components of power, respectively.

From Antoniu,[9] the expression for the per-phase apparent rms power, S_e, as adapted in Arrillaga and Watson[6] is:

$$S_e = \sum_k \sqrt{\left(P_k^2 + Q f_k^2\right)} = \sum_k V_k I_k \qquad (1.60)$$

and the apparent power for a three-phase system, S_s:

$$S_s = V_{rms} I_{rms} = \sqrt{P^2 + Q_f^2} = \sqrt{\sum_k V_k^2} \sqrt{\sum_k I_k^2} \qquad (1.61)$$

where Q_f is the reactive power.

Emanuel[12] is an advocate for the separation of power in fundamental and nonfundamental components and further proposes the determination of apparent power, S, as:

$$S = \sqrt{S_1^2 + S_n^2} \qquad (1.62)$$

where S_1 is the fundamental and S_n the nth component of apparent power. The harmonic active power, P_H, embedded in S_n is negligible, around half a percent of the fundamental active power, according to Kusters and Moore.[11]

1.6.7 Voltage in Balanced Three-Phase Systems

Harmonics of different order form the following sequence set:

Fundamentals of Harmonic Distortion and Power Quality Indices

Positive sequence: 1, 4, 7, 10, 13, …
Negative sequence: 2, 5, 8, 11, 14, …
Zero sequence: 3, 6, 9, 12, 15, … (also called triplen)

The positive sequence system has phase order R, S, T (a, b, c) and the negative sequence system has phase order R, T, S (a, c, b). In the zero sequence system, the three phases have an equal phase angle. This results in a shift for the harmonics, which for a balanced system can be expressed as follows:

$$Va_h(t) = \sqrt{2}V_h \sin(h\omega_0 t + \theta_h) \tag{1.63}$$

$$Vb_h(t) = \sqrt{2}V_h \sin(h\omega_0 t - \frac{2h\pi}{3} + \theta_h) \tag{1.64}$$

$$Vc_h(t) = \sqrt{2}V_h \sin(h\omega_0 t + \frac{2h\pi}{3} + \theta_h) \tag{1.65}$$

Note that in Equation (1.60) through Equation (1.62), harmonics voltages and currents are displaced 120° from one another. The phase sequences of harmonic voltage or currents can be described as in Table 1.1. Notice that triplen harmonics are zero sequence.

In an unbalanced system, harmonic currents will contain phase sequences different from those in Table 1.1.

1.6.8 Voltage in Unbalanced Three-Phase Systems

Unbalanced voltage conditions are rare but possible to find in three-phase electric power systems. The main reason for voltage unbalance is an irregular distribution of single-phase loads; other reasons may include mutual effects in asymmetrical conductor configurations. During load or power system unbalance, it is possible to find voltages of any sequence component:

$$\begin{bmatrix} V1_h \\ V2_h \\ V3_h \end{bmatrix} = \frac{1}{3} \begin{bmatrix} 1 & a & a^2 \\ 1 & a^2 & a \\ 1 & 1 & 1 \end{bmatrix} \begin{bmatrix} Va_h \\ Vb_h \\ Vc_h \end{bmatrix} \tag{1.66}$$

where $a = e^{j120°}$.

In most cases, there is a dominant sequence component with a meager contribution from other frequencies. Under certain conditions involving triplen harmonics, there can be only positive or negative sequence components.

TABLE 1.1
Phase Sequences of Harmonics in a Three-Phase Balanced System

Harmonic order	1	2	3	4	5	6	7	8	9	10	11 …
Phase sequence	Positive	Negative	Zero	Positive	Negative	Zero	Positive	Negative	Zero	Positive	Negative …

REFERENCES

1. De Almeida, A., Understanding Power Quality, *Home Energy Magazine Online*, November/December 1993, http://homeenergy.org/archive/hem.dis.anl.gov/eehem/93/931113.html.
2. Edminster, J.A., *Electrical Circuits*, McGraw Hill, Schaum's Series, New York, 1969.
3. IEC 61000-4-7 Edition 2, Electromagnetic compatibility (EMC) — part 4-7: testing and measurement techniques — general guide on harmonics and interharmonics measurements and instrumentation, for power supply systems and equipment connected thereto, 2002.
4. Electrotek Concepts, TOP — The Output Processor, http://www.pqsoft.com/top/.
5. IEEE Power Engineering Society, *IEEE Tutorial on Modeling and Simulations*, IEEE PES, 1998.
6. Arrillaga, J. and Watson, N., *Power Systems Harmonics*, 2nd ed., Wiley, New York, 2003.
7. IEEE Std. 519-1992 Recommended Practices and Requirements for Harmonic Control in Electric Power Systems.
8. Frank, J.M., Origin, development and design of K-factor transformers, *IEEE Ind. Appl. Mag.*, Sept/Oct 1997.
9. Antoniu, S., Le régime energique deformant. Une question de priorité, *RGE*, 6/84, 357–362, 1984.
10. Fryze, S., Wirk, Blind und Scheinleistung in Electrischen Stromkreisien mit Nitchsinuformigen Verlauf von Strom und Spannung, *Electrotechnisch Zeitschrift*, 596–599, June, 1932.
11. Kusters, N.L. and Moore, W.J.M., On definition of reactive power under non sinusoidal conditions, *IEEE Trans. Power Appar. Syst.*, 99, 1845–1850, 1980.
12. Emanuel, A.E., Power in nonsinusiodal situations, a review of definitions and physical meaning, *IEEE Trans. Power Delivery*, 5, 1377–1383, 1990.
13. Emanuel, A.E., Apparent power components and physical interpretation, *Int. Conf. Harmonics Qual. Power (ICHQP'98)*, Athens, 1998, 1–13.

2 Harmonic Sources

2.1 INTRODUCTION

Although power system harmonics is the topic of this book, it is important to stress that harmonic waveform distortion is just one of many different disturbances that perturb the operation of electrical systems. It is also a unique problem in light of an increasing use of power electronics that basically operate through electronic switching. Fortunately, the sources of harmonic currents seem to be sufficiently well identified, so industrial, commercial, and residential facilities are exposed to well-known patterns of waveform distortion.

Different nonlinear loads produce different but identifiable harmonic spectra. This makes the task of pinpointing possible culprits of harmonic distortion more tangible. Utilities and users of electric power have to become familiar with the signatures of different waveform distortions produced by specific harmonic sources. This will facilitate the establishment of better methods to confine and remove them at the sites where they are produced. In doing this, their penetration in the electrical system affecting adjacent installations will be reduced. As described in Chapter 6 and Chapter 8, parallel resonant peaks must be properly accounted for when assessing waveform distortion. Otherwise, a filtering action using single–tuned filters to eliminate a characteristic harmonic at a given site may amplify the waveform distortion if the parallel peak (pole) of the filter coincides with a lower order characteristic harmonic of the load. Active filters may overcome this hurdle but they must be well justified to offset their higher cost.

The assessment of harmonic propagation in a distribution network, on the other hand, requires an accurate representation of the utility source. Weak sources will be associated with significant harmonic distortion that can in turn affect large numbers of users served from the same feeder that provides power to the harmonic-producing customer. This will become particularly troublesome when harmonics are created at more than one location — for example, in a cluster of industrial facilities served from the same feeder. Thus, utilities may be inadvertently degrading the quality of power by serving heavy harmonic producers from a weak feeder.

From the perspective of the customer, power quality means receiving a clean sinusoidal voltage waveform with rms variations and total harmonic distortion within thresholds dictated by a number of industrial standards. Often, however, utilities find it difficult to keep up with these regulations. The culprit is often found on the customer loads, which from victims they turn into offenders when they draw large blocks of currents from the AC source in slices. This occurs whenever they convert power from one form into another through rectification and inversion processes. The waveform chopping process leads to noise-like structures often regarded as "dirty," "unclean," or "polluted" power. This is nothing but the harmonic distortion of the voltage supply, which is the subject matter of this book and must be assessed from all possible perspectives.

As it will be described, the main effects of this distortion range from increased equipment losses that shorten the lifetime of equipment like transformers and cables to interference in audio and data communication and to possible protective devices' nuisance tripping. The cost involved in cleaning this harmonic noise will often counteract the benefit obtained from improving equipment and appliances to better operate them in disturbed environments. As a general rule, the more sophisticated or sensitive electronic equipment is, the higher the cost to keep it running given its increased sensitivity to power quality disturbances. Sensitive industrial processes, such as highly automated assembly lines, are prone to power-related damage from severe harmonic distortion. On the other hand, the stiffer the AC source is, the lower the voltage distortion across the source terminals will be.

Harmonic distortion is no longer a phenomenon confined to industrial equipment and processes, where the first power quality concerns developed. Uninterruptible power supplies (UPSs), personal computers (PCs), and electronic and entertaining devices proliferate nowadays in commercial and residential installations. These special kinds of loads represent formidable sources of harmonic currents and they increase with the expanding use of video recorders, digital clocks, and other sensitive electronic equipment.

The interaction between power utility AC sources and customer loads will be further affected by distributed resources (often known as distributed generators or DGs) that employ electronic switching technologies, increasingly used by utilities and industry to better cope with peak demand. The expanding presence of this type of DGs in distribution networks will contribute to an overall rise in harmonic distortion in electric distribution networks. An increased effort by utilities and industries alike to discover improved mitigation methods that can keep harmonic distortion within allowable limits is foreseen.

2.2 THE SIGNATURE OF HARMONIC DISTORTION

Figure 2.1 illustrates a simple PSCAD (Power Systems Computer-Aided Design)[1] model to produce distortion on the voltage waveform. The student edition of this software has been used for this and other examples presented throughout the book.[1] Consequently, only simple cases in which modeling can be achieved with the reduced number of nodes available are considered in the examples. For larger applications, the student edition falls short of dimensions to model all relevant features of an electrical installation and complex loads. Though simplified, the model in Figure 2.1 provides a practical glance at the effects of harmonic currents of different frequencies on AC voltage waveform signatures. The harmonic injected current was kept constant and the simulation included harmonic currents of the 2nd to the 11th order.

Figure 2.2(a) through Figure 2.2(j) show the corresponding results. Even (not odd) harmonics are unlikely to exist in balanced systems; however, in this example, we include them to get an overall idea of the different waveform distortion signatures.

FIGURE 2.1 Harmonic generator to determine the voltage waveform distortion at different harmonic frequencies.

Even harmonics may arise (for instance, under waveform asymmetry) if thyristor triggering angles were slightly different on every half cycle. This is by no means a comprehensive assortment of all the harmonic distortion signatures likely to be found in practical situations. Just by varying the phase angle and amplitude of the harmonic currents relative to the fundamental frequency, we would assemble an endless collection of different distorted waveforms.

2.3 TRADITIONAL HARMONIC SOURCES

Prior to the development of power electronic switching devices, harmonic current propagation was looked at from the perspective of design and operation of power apparatus devices with magnetic iron cores, like electric machines and transformers. In fact, at that time the main source of harmonics must have involved substation and customer transformers operating in the saturation region.

Nowadays, harmonic distortion produced under transformer saturation probably at peak demand or under elevated voltage during very light load conditions is only one of numerous situations that generate harmonic waveform distortion. Possibly, electric furnaces should be regarded as the second most important cause of concern in high-power applications in industry, second to power converter utilization.

The sources of waveform distortion in power systems are multiple and, in industrial installations, they can be found from small (less than 1 kVA) to several tens of megavoltamperes. However, as mentioned earlier, commercial and residential facilities can also become significant sources of harmonics. This is particularly true when the combined effects of all individual loads served by the same feeder are taken into account. For instance, a simple power source of a home desktop computer may draw around 4 A from a 127-V main, or around 500 VA. A medium voltage feeder typically serving around 2500 LV customers would be eventually supplying around 1.25 MVA of computer power under the likely scenario of having all customers checking e-mail accounts or browsing in the Internet in unison after dinner.

A six-pulse converter shows a theoretical amplitude (as shortly addressed in this chapter) of around 20% just for the fifth harmonic current, as it was depicted in Figure 1.1. This translates into a similar percentage of reactive power. In our example, this amounts to around 1.25 × 0.20 or 250 kVA of 300 Hz power that can be

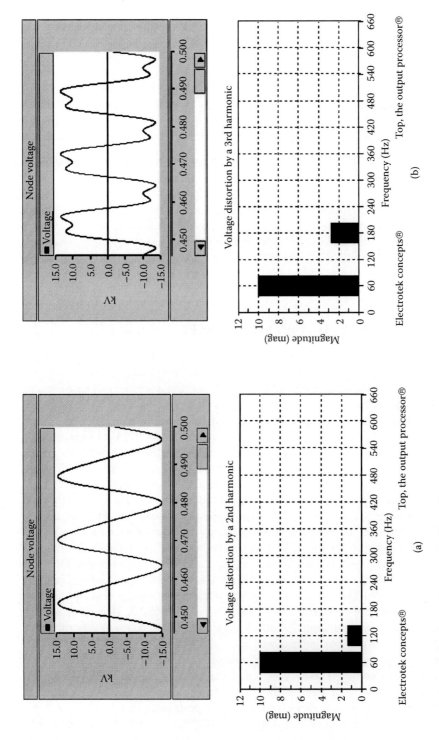

FIGURE 2.2 Waveform distortion imposed by currents of harmonic orders 2 through 11.

Harmonic Sources

FIGURE 2.2 *Continued.*

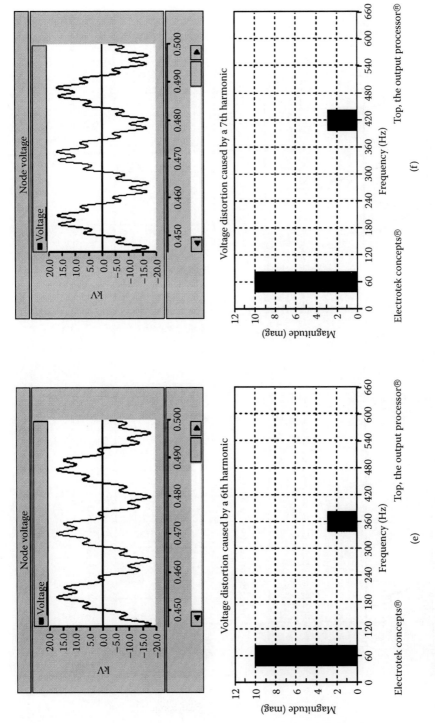

FIGURE 2.2 Continued.

Harmonic Sources

FIGURE 2.2 *Continued.*

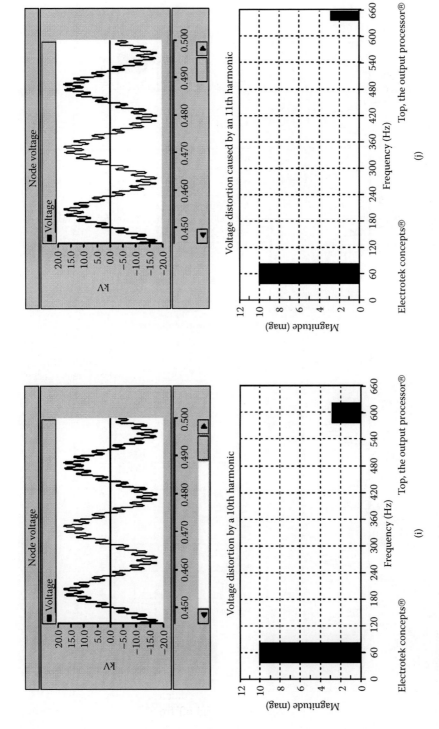

FIGURE 2.2 Continued.

envisaged as current injection back into the power system. This contributes to the distortion of the voltage waveform. Other harmonics generated by the power converter would add additional distortion to the voltage waveform because they can be conceived as individual spectral currents injected into the power system.

As previously mentioned, some harmonic sources like saturated transformers have existed from the early times when electricity was first transformed and distributed over power lines for commercial purposes. As described in the next section, the operation of transformers near the saturation zone is the result of excessive magnetic flux through the core, which limits the linear increase of the magnetic flux density. Rotating machines are another example of equipment that may behave as a harmonic distortion source under overloading conditions.

The use of electricity involving loads that require some form of power conditioning like rectification and/or inversion is on the rise, as mentioned in Chapter 1. The greatest majority of industrial nonlinear loads are related to solid-state switching devices used in power converters that change electric power from one form to another. This includes, among others, AC to DC energy conversion for DC motor speed control, and AC to DC and back to AC at variable frequencies for processes involving speed control of induction motors. Most bulk energy conversion processes take place in the oil, mining, steel mill, pulp and paper, textile, and automobile industries. Other applications include manufacturing assembly lines and electrolytic coating processes, which can produce significant amounts of harmonic current generation.

Arc devices (namely, electric furnaces, soldering equipment, fluorescent and mercury-vapor or high-pressure sodium lamps) can become very special sources of harmonics in that they can involve frequencies below the fundamental power frequency and fractional harmonics. The former are called subharmonics and the latter are known as interharmonics. Subharmonic generation can typically take place when arc-type devices are sourced through weak transmission or distribution systems, i.e., those with small ratios of short circuit to load current.

Large inrush currents during switching of capacitor banks, transformers, and rotating machines into the distribution system can develop harmonic currents. IEEE-519[2] allows for harmonic distortion limits 50% higher than recommended values during start-ups and unusual conditions lasting less than 1 hour. Harmonic distortion due to inrush currents on transformer energization and to outrush currents developed under shunt connection of capacitor banks (especially when more capacitors are added to an existing bank) fall in this category. In the latter case, large currents are discharged from one bank into the other because the only limiting element existing between the two banks is the surge impedance of the connecting lead. This may explain some nuisance operations of capacitor bank fuses.

Other harmonic sources may include ferroresonance phenomena, which may remain undetected for periods of minutes and even hours until reported by affected customers. Thus, the severe waveform distortion developed no longer fits into the IEEE-519 definition for unusual conditions. This undesired event has the potential to produce extensive damage to customer facilities and equipment and therefore, utilities and industry must minimize the risk to trigger this condition. American[3] and European[4] publications address this phenomenon and provide guidance to assess the possibility of ferroresonance involving small transformers fed off from underground cables.

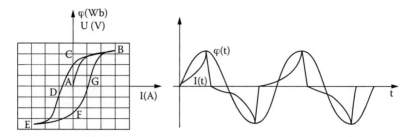

FIGURE 2.3 Distorted current under transformer saturation conditions.

2.3.1 Transformers

A transformer can incur in core saturation conditions in either of the following cases:

When operating above rated power
When operating above rated voltage

The first situation can arise during peak demand periods, and the second case can occur during light load conditions, especially if utility capacitor banks are not disconnected accordingly and the feeder voltage rises above nominal values.

A transformer operating on the saturation region will show a nonlinear magnetizing current similar to that illustrated in Figure 2.3, which contains a variety of odd harmonics, with the third dominant. The effect will become more evident with increasing loading. In an ideal lossless core, no hysteresis losses are produced. The magnetic flux and the current needed to produce them are related through the magnetizing current of the steel sheet material used in the core construction. Even under this condition, if we plot the magnetizing current vs. time for every flux value considered, the resultant current waveform would be far from sinusoidal.

When the hysteresis effect is considered, this nonsinusoidal magnetizing current is not symmetrical with respect to its maximum value. The distortion is typically due to triplen harmonics (odd multiples of three, namely, the 3rd, 9th, 15th, etc.), but mainly due to the third harmonic. This spectral component can be confined within the transformer using delta transformer connections. This will help maintain a supply voltage with a reasonable sinusoidal waveform.

In three-legged transformers, the magneto motive forces (mmf) of triplen harmonics are all in phase and act on every leg in the same direction. Therefore, the trajectory of the magnetic flux for the triplen harmonics extends outside the boundaries of the core. The high reluctance of this trajectory reduces the flux of triplen harmonics to a very small value. The components of fifth and seventh harmonics can also be considerable (5 to 10%) to produce considerable distortion and ought not be ignored.

In electric power distribution networks, harmonics due to transformer magnetizing current reach their maximum value early before dawn when the system is lightly loaded and voltage level is high. When a transformer is de-energized, it is possible that it retains residual magnetic flux in the core. On re-energization, this flux coalesces with the magnetizing flux produced by the inrush current, and the two combined can yield peak values three times or higher the nominal flux at rated load. The resulting effect may cause the transformer core to reach extreme saturation levels involving excessive turn amps within the core. Consequently, magnetizing currents as large as 5 to 10 p.u. of nominal current (compared with 1 to 2% of nominal magnetizing current during steady-state operating conditions) can develop. The duration of the magnetizing current is mainly a function of the primary winding resistance. For large transformers with large winding resistance, this current can remain for many seconds.

The harmonic content of steady-state currents in three-phase systems does not involve even harmonics, which appear under waveform asymmetry when the positive and negative half cycles are not of the same amplitude. However, under energization, a distribution transformer develops all kinds of low-order harmonics involving even harmonics (notably the second and the fourth, as depicted in Figure 2.4), which are often used for restraining the operation of differential protection.

The transformer model used in the PSCAD software is the so-called unified magnetic equivalent circuit transformer or UMEC,[5,6] which overcomes the problem of pulling together the information on transformer core and winding characteristics by deriving the elements of the inductance matrix from test data on open- and short-circuit tests. UMEC simulates the nonlinear characteristic of the core by using a piecewise linear representation of the equivalent branch conductance, as shown in Figure 2.5. This allows reducing the processing time by shortening the number of matrix inversions.

The presence of the even harmonics and their decaying nature are typical under transformer saturation. Here we can reproduce the harmonics created during transformer saturation by injecting harmonic currents similar to the typical harmonic spectrum of Figure 2.4 into an AC source and a short feeder representation. Figure 2.6 depicts the simplified model recreated in PSCAD. The load voltage and current waveforms and their harmonic spectra are shown in Figure 2.7 for a 180° phase angle between fundamental and harmonic currents.

Resultant THD values for source voltage (V_1) and load current (I_{total}) are 6 and 66%, respectively. Note on the bottom plot the distorted voltage waveform of the load relative to the voltage waveform of the source, which also undergoes some distortion. Figure 2.8 shows corresponding results for a weak AC source with impedance around 50% larger than that considered in generating the results in Figure 2.7. In the latter case, THD values resulted in 9 and 98% for voltage and current, respectively, *a substantial increase*. This illustrates the important role that the source impedance plays is in determining the voltage distortion levels.

As noticed, voltage as well as current waveforms suffer from distortion produced by harmonic currents during transformer saturation phenomena. Transformer saturation can also take place following a voltage dip because a sudden change in voltage leads to a DC component in the magnetizing flux.

2.3.2 ROTATING MACHINES

As a result of small asymmetries on the machine stator or rotor slots or slight irregularities in the winding patterns of a three-phase winding of a rotating machine,

FIGURE 2.4 Typical transformer energizing current.

harmonic currents can develop. These harmonics induce an electromotive force (emf) on the stator windings at a frequency equal to the ratio of speed/wavelength. The resultant distribution of magneto motive forces (mmfs) in the machine produces harmonics that are a function of speed. Additional harmonic currents can be created upon magnetic core saturation. However, these harmonic currents are usually smaller than those developed when the machines are fed through variable frequency drives (VFDs).

Additional discussion on harmonics in rotating machines is provided in Chapter 4.

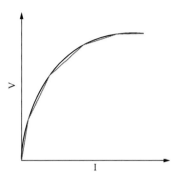

FIGURE 2.5 Piecewise linear representation of transformer conductance.

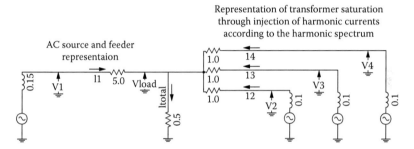

FIGURE 2.6 Simplified way to simulate harmonic injection into an AC source, typical of transformer saturation.

2.3.3 POWER CONVERTERS

The increasing use of the power conditioners in which parameters like voltage and frequency are varied to adapt to specific industrial and commercial processes has made power converters the most widespread source of harmonics in distribution systems. Electronic switching helps the task to rectify 50-/60-Hz AC into DC power. In DC applications, the voltage is varied through adjusting the firing angle of the electronic switching device. Basically, in the rectifying process, current is allowed to pass through semiconductor devices during only a fraction of the fundamental frequency cycle, for which power converters are often regarded as energy-saving devices. If energy is to be used as AC but at a different frequency, the DC output from the converter is passed through an electronic switching inverter that brings the DC power back to AC.

Converters can be grouped into the following categories:

Large power converters like those used in the metal smelter industry and in HVDC transmission systems

Medium-size power converters like those used in the manufacturing industry for motor speed control and in the railway industry

Small power rectifiers used in residential entertaining devices, including TV sets and personal computers. Battery chargers are another example of small power converters.

FIGURE 2.7 Waveform distortion due to harmonic currents, typical of transformer saturation.

Harmonic Sources

FIGURE 2.8 Waveform distortion due to harmonic currents, typical of transformer saturation in a weaker system.

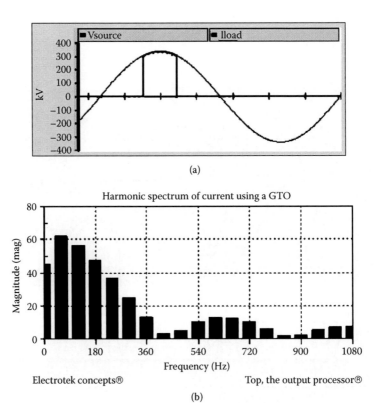

FIGURE 2.9 AC switch using a GTO and harmonic spectrum of the current.

Figure 2.9(a) describes the basic relation between current and voltage in which a half-way control using a GTO (gated turn-off thyristor, as portrayed by Finney[7]) to draw current during part of the AC waveform positive half cycle. A strong DC component due to the switching action taking place only on one side of the AC cycle is manifest in Figure 2.9(b). In addition, zero sequence (triplen) and even harmonic components are evident. This peculiar behavior is also characteristic of unbalanced three-phase systems.

If a different switching device like an IGBT (insulated gate bipolar transistor, as described by Finney[7]) in which current flows during only part of the time on every half cycle (Figure 2.10a) is used, the current harmonic spectrum of Figure 2.10(b) is obtained.

Notice how, in this case, the waveform does not contain even harmonics due to the symmetry of the switching action relative to the x-axis. Therefore, only odd and zero sequence harmonics show up. The same result can be achieved through Fourier analysis decomposition of any waveform containing identical features on the two half cycles.

To further illustrate the power converter as a harmonic source, let us refer to the six-pulse VFD of Figure 2.11. Harmonic currents, i_{hL}, produced by the rotating machine will be practically confined to the load side, beyond the DC bus. This is regardless of whether the converter is of a current source (a) or a voltage source (b) configuration. Due to the commutation of current from one phase to another during

Harmonic Sources

FIGURE 2.10 AC switch using an IGBT and harmonic spectrum of the current.

the rectifying process on the converter, harmonic currents, i_{hS}, will show up on the source side. It can be shown that the current drawn by the six-pulse bridge contains harmonics of the order:

$$n = (P * i \pm 1) \tag{2.1}$$

where i is an integer greater than or equal to 1.

For a six-pulse converter, P = 6 and the line current contains harmonics of the order 5, 7, 11, 13, ... These are referred to as the characteristic harmonics of the six-pulse converter. For a 12-pulse converter (two six-pulse units in parallel), its characteristic harmonics will be: 11, 13, 23, 25, 35, 37, ... This is the reason behind the common idea that one way to control harmonics is to use converters with higher numbers of pulses.

For a six-pulse converter, the following observations apply:

No triplen harmonics are present.
There are harmonics of order $6k \pm 1$ for integer values of k.
Harmonics of order $6k + 1$ are of positive sequence.
Harmonics of order $6k - 1$ are of negative sequence.

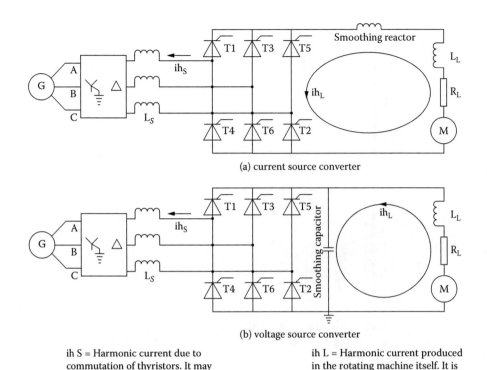

FIGURE 2.11 Six-pulse converter used as a DC motor speed controller.

Twelve-pulse converters are powered from a three-winding (or phase shift) transformer, with a phase difference of 30° between the secondary and the tertiary; each connects to a converter's bridge. These converters create harmonics of order $12k\ (\pm)\ 1$ at the source side. The harmonic currents of order $6k\ (\pm)\ 1$ (with k odd), i.e., $k = 5, 7, 17, 19$, etc., flow between the secondary and tertiary of the phase shift transformer but do not make their way into the AC network.

The amplitude of the harmonic current on the converter front end will be influenced by the presence of a smoothing reactor such as that shown in Figure 2.11(a). For a six-pulse diode bridge having a large smoothing reactor, the magnitude of the harmonics can be approximated by the expression:

$$|I_h| \approx \frac{|I_{fund}|}{h} \qquad (2.2)$$

where $|I_h|$ and $|I_{fund}|$ are the magnitude of the nth-order harmonic and the fundamental current, respectively.

Higher harmonic currents can be expected if the smoothing reactor is small or nonexistent (see Figure 2.11(b). Triplen harmonics can appear under unbalanced volt-

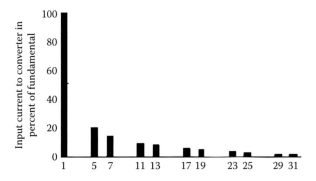

FIGURE 2.12 Harmonic spectrum of a six-pulse converter.

ages. For this bridge, the harmonic spectrum will thus look like that in Figure 2.12. Note that the fifth harmonic shows a value of 20% as obtained from Equation (2.2).

Fundamental mmf rotates in the positive direction, mmf from triplen harmonics is absent, and any fifth and seventh harmonic mmfs rotate in the negative and positive directions, respectively. Thus, from looking at the spectrum of Figure 2.12, it is possible to understand that negative sequence torques (from 5th, 11th, 17th, etc., harmonics) will be strongly interacting with positive sequence torques (from 7th, 13th, 19th, etc., harmonics) to produce torsional pulsating torques. This may explain the increased vibration levels sometimes experienced in applications involving synchronous generators feeding large VFDs industrial applications, as further discussed in Chapter 4.

From Alex McEachern's Teaching Toy Edition 2.0 (a useful and educational free harmonics tool particularly suitable for students),[8] a three-phase bridge rectifier would reveal the current waveform depicted in Figure 2.13.

2.3.3.1 Large Power Converters

These are used in electric utility applications in which large blocks of energy are transformed from AC to DC. Their nominal power is in the megavoltampere range and, generally, they present a much higher inductance on the DC than on the AC side. Therefore, the DC current is practically constant and the converter acts as a harmonic voltage source on the DC side and as a harmonic current source on the AC side. Furthermore, in a perfectly balanced system, all resultant currents are the same in all phases.

2.3.3.2 Medium-Size Power Converters

Kilovoltampere-sized converters are in this range and are found in increasing numbers in industry. The first applications in the industry were for DC motor speed control that still represents the major market for these types of converters. However, they are increasingly used in AC induction motor speed control. Many applications are now seen at land and offshore applications in the oil industry, where submersible pumping systems are used with variable frequency drives as artificial methods for

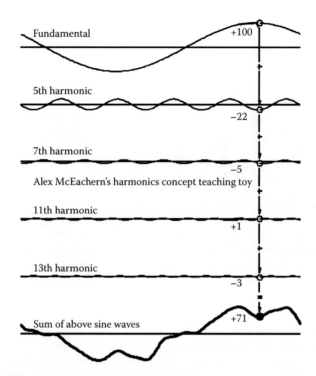

FIGURE 2.13 Three-phase rectifier. (Adapted from McEachern, A., Power Quality Teaching Toy, Edition 2.0, 2005.)

oil production. Furthermore, the advent of power transistors and GTO thyristors is progressively stimulating the use of power converters for AC motor speed control.

Similar to large-size power converters, the fifth harmonic can reach amplitudes that range from one fifth to one third the fundamental rated current.

In the case of electric railroad applications, it is common to see individual controls in every rectifier bridge. During the initial accelerating period with maximum current in the DC motor, the rectifier bridge produces the worst harmonic currents and operates at a low power factor. To alleviate this condition at low speeds, one of the bridges is bypassed while phase control is applied to the other bridge. Table 2.1 recaps the different applications of medium-size power converters.

2.3.3.3 Low-Power Converters

Uninterruptible power supplies (UPSs), welders, and printers are among these low-kilovoltampere–size power converter applications. It is common to see large commercial and public office buildings stuffed with computers and other peripheral devices. If they are additionally provided with UPSs to handle voltage sags and power supply interruptions, the amounts of harmonic currents can substantially increase. Residential areas at specific times of the day act as fabulous harmonic sources produced by all kinds of entertaining devices, as described previously.

Harmonic Sources

TABLE 2.1
Some Power Converter Applications in Industry

Manufacturing Industry	Paper and Steel Industries	Transportation, Textile, and Food Industries	Petrochemical Industry	Residential Installations
Agitators, pumps, fans, and compressors in the process industry; planers, winches, drill presses, sanders, extruders, grinders, mills, and presses in machining	Blowers and compressors in heating and air conditioning; hoists and steel mill rollers	Elevators, trains, automobiles in transportation; looms in textiles; sheers in packaging; conveyors and fans in food industries	Compressors, variable frequency drives to power oil pumps, fans, cranes, and shovels in the oil and gas installations	Heat pumps, freezers, and washing machines

The individual harmonics generated by battery charger circuits depend on the initial battery voltage. The overall harmonic content varies as a function of time and involves a random probability.

As in other appliances that use DC current (TV sets, radio and stereo amplifiers, etc.) battery chargers produce zero sequence harmonics, which overload the neutral conductor of the three-phase distribution transformer that supplies the single-phase, low-voltage loads. This is because the phase angle of the third harmonic does not vary enough to produce harmonic cancellation, so they are added up algebraically. As later discussed, fluorescent lighting also produces triplen harmonics, for which a concurrent use of battery chargers and fluorescent lamps from the same circuit can make things even worse.

Unlike the types of loads described earlier, which nominal power is large enough to deserve an individual treatment, the loads we refer to in this section are important only when they represent a significant portion of the total load under concurrent operation. Monte Carlo method can be used in some applications to investigate the probability of exceeding preset levels of harmonics from TV sets as well as from electric vehicle battery chargers serving multiple locations within the network.

2.3.3.4 Variable Frequency Drives

VFDs are, in reality, power converters. The reason to further address them under a separate section is because, by themselves, VFDs constitute a broad area of application used in diverse and multiple industrial processes. In a very general context, two types of VFDs can be distinguished: those that rectify AC power and convert it back into AC power at variable frequency and those that rectify AC power and directly feed it to DC motors in a number of industrial applications.

In both cases, the front-end rectifier, which can make use of diodes, thyristors, IGBTs, or any other semiconductor switch, carry out the commutation process in which current is transferred from one phase to the other. This demand of current "in slices" produces significant current distortion and voltage notching right on the source side, i.e., at the point of common coupling. Motor speed variations, which are achieved through firing angle control, will provide different levels of harmonic content on the current and voltage waveforms.

Variable frequency drive designs also determine where harmonic currents will predominantly have an impact. For example, voltage source inverters produce complex waveforms showing significant harmonic distortion on the voltage and less on the current waveforms. On the other hand, current source inverters produce current waveforms with considerable harmonic contents with voltage waveforms closer to sinusoidal. None of the drive systems is expected to show large distortion on both voltage and current waveforms, in line with Finney's observations.[7]

2.3.3.4.1 Distribution Static Compensator (DSTATCOM)

The DSTATCOM is a good example of a voltage source inverter (VSI) power electronics device connected in shunt to the distribution network. This is a concept imported from the application of FACTS (flexible AC transmission systems) comprehensively described by Hingorani and Gyugyi.[9] Among the objectives of the DSTATCOM are to eliminate harmonics from the power supply and to provide voltage and reactive power support during faults in the system. However, because the DSTATCOM uses a rectification bridge, a continuous harmonic production is created on the source side. Thus, the example presented here illustrates the harmonic voltage distortion in a circuit involving a distribution static compensator during and after a three phase to ground fault.

The fault is simulated to occur at $t = 1.5$ s and last 0.75 s. The example, which is modeled using the PSCAD student edition software, involves a voltage control with PI (Proportional-Integral) controller and a pulse width modulation (PWM) controller, with carrier frequency of nine times fundamental and varying DC voltage. Figure 2.14 depicts the diagram of the six-pulse STATCOM set-up and Figure 2.15 shows the calculated voltage waveforms at both ends of the rectifying bridge in a time window that encompasses the start and end of the fault.

Figure 2.16 presents results of the simulation, which shows harmonic distortion and spectral content at the onset and extinction of the staged fault.

The THD_V levels found for voltage, Vna, during the transition times are as follows:

1.5 to 1.6 s: $THD_V = 12.77\%$ (fault starting period)
2.25 to 2.3 s: $THD_V = 13.95\%$ (fault clearing period)
2.5 s onward (until 2.5 s): $THD_V = 5\%$ (postfault or steady state)

Notice that the total harmonic distortion levels during and at the clearing periods of the fault are more than twice the steady-state levels. At first glance, these levels are above the recommended standard values described in Chapter 3. However, standards do not cover harmonic distortion during transient conditions or during short-circuit faults. The fault example is chosen here to illustrate the ability of the software to calculate harmonic distribution change in rapid succesion.

Harmonic Sources

FIGURE 2.14 Six-pulse STATCOM.

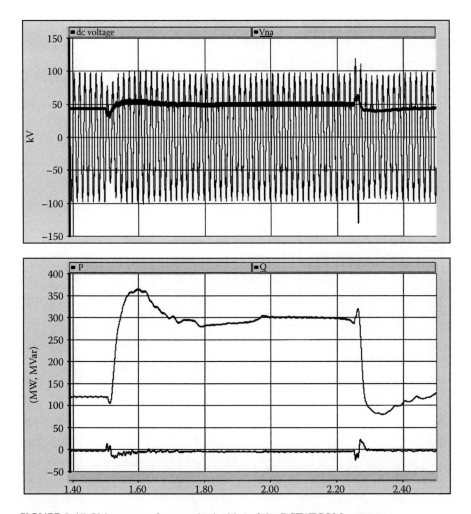

FIGURE 2.15 Voltage waveforms at both sides of the DSTATCOM converter.

Harmonic Sources

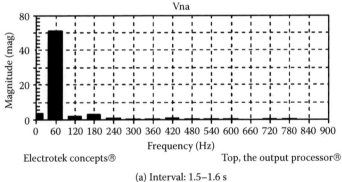

(a) Interval: 1.5–1.6 s

FIGURE 2.16 Harmonic spectra of the PWM voltage control system of Figure 2.15.
Continued.

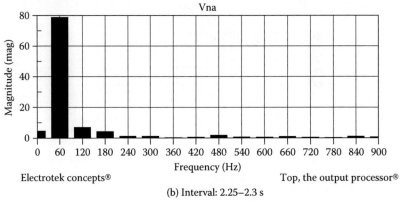

Electrotek concepts® Top, the output processor®
(b) Interval: 2.25–2.3 s

FIGURE 2.16 *Continued.*

Harmonic Sources

Electrotek concepts® Top, the output processor®

(c) Interval 2.3–2.5 s

FIGURE 2.16 *Continued.*

(d) Summary of harmonic spectra including commencement, end– and postfault periods, respectively

FIGURE 2.16 *Continued.*

2.3.4 Fluorescent Lamps

Fluorescent tubes are highly nonlinear in their operation and give rise to odd harmonic currents of important magnitude. As a brief portrayal of the fluorescent lamp operation, we can state that magnetic core inductors or chokes contained inside the start ballasts function to limit the current to the tube. Likewise, they use a capacitor to increase the efficiency of the ballast by increasing its power factor. Electronic ballasts operate at higher frequency, which permits the use of smaller reactors and capacitors. The use of higher frequencies allows them to create more light for the same power input. This is advantageously used to reduce the input power.

In a four-wire, three-phase load, the dominant phase current harmonics of fluorescent lighting are the third, fifth, and seventh if they use a magnetic ballast and the fifth with an electronic ballast (as adapted from Tollbert et al.[10] and presented in Figure 2.17). Triplen harmonics are added in the neutral, being the third the dominant for a magnetic ballast but multiple harmonics if an electronic ballast is used. See Figure 2.18.

In Figure 2.18, notice the impressive amount of third-order harmonics in neutral, particularly because they are all added in phase. It will be nonetheless important to remember that the current in neutral must be determined from:

$$I_{neutral} = \sqrt{I_1^2 + I_3^2 + I_5^2 + I_7^2 + I_9^2 + I_{11}^2 + \ldots + I_n^2} \qquad (2.3)$$

Harmonic Sources

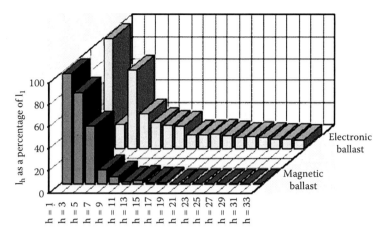

FIGURE 2.17 Harmonic spectra of fluorescent lamps for phase currents. (Data from Tollbert et al.)

FIGURE 2.18 Harmonic spectra of fluorescent lamps for neutral currents. (Data from Tollbert et al.)

Furthermore, lighting circuits frequently involve long distances and combine with a poorly diversified load. With individual power factor correction capacitors, the complex LC circuit can approach a resonant condition around the third harmonic. Therefore, these are significant enough reasons to oversize neutral wire lead connections in transformers that feed installations with substantial amounts of fluorescent lighting. Capacitor banks may be located adjacent to other loads and not necessarily as individual power factor compensators at every lamp.

2.3.5 Electric Furnaces

The melting process in industrial electric furnaces is known to produce substantial amounts of harmonic distortion. The introduction of fundamental frequency harmonics develops from a combination of the delay in the ignition of the electric arc along with its highly nonlinear voltage-current character. Additionally, voltage changes caused by the random variations of the arc give rise to a series of frequency variations in the range 0.1 to 30 kHz; each has its associated harmonics. This effect is more evident in the melting phase during the interaction of the electromagnetic forces among the arcs. Figure 2.19 shows plots of (a) the electric furnace power and (b) the current harmonics without attenuation of harmonic filters in a typical electric furnace application. The example includes the application of the Smart Predictive Line Controller, a patented Hatch technology[11] for arc stabilization and flicker reduction on AC electric arc furnaces.

2.4 FUTURE SOURCES OF HARMONICS

The challenge for electrical system designers in utilities and industry is to design the new systems and/or adapt the present systems to operate in environments with escalating harmonic levels. The sources of harmonics in the electrical system of the future will be diverse and more numerous. The problem grows complicated with the increased use of sensitive electronics in industrial automated processes, personal computers, digital communications, and multimedia.

Utilities, who generally are not regarded as large generators of harmonics, may be lining up to join current harmonic producers with the integration of distributed resources in the rise. Photovoltaic, wind, natural gas, carbonate full cells, and even hydrogen are expected to play increasingly important roles in managing the electricity needs of the future. Distributed generators that presently provide support to utilities, especially during peak demand hours, will be joined by numerous harmonic-producing units, fueled by natural gas or even wind, called microturbines.

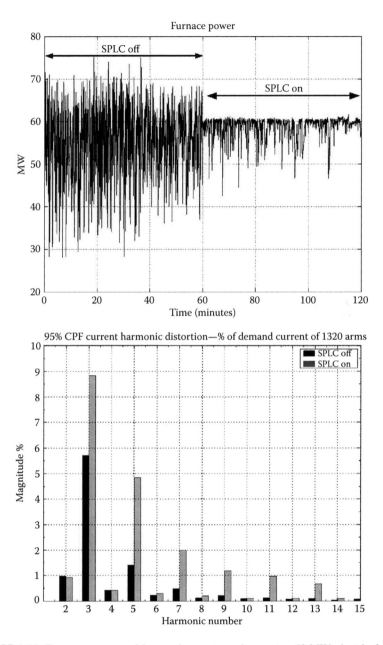

FIGURE 2.19 Furnace power and harmonic spectrum drawn at a 60-MW electric furnace by a typical arc furnace. (Adapted from private communication from Hatch, Smart Predictive Line Controllers, 2005.)

REFERENCES

1. Power System Computer Aided Design (PSCAD), http://pscad.com.
2. IEEE Std. 519-1992 IEEE Recommended Practices and Requirements for Harmonic Control in Electric Power Systems.
3. IEEE WG on modeling and analysis of system transients using digital programs, *IEEE Trans. Power Delivery*, 15(1), Jan 2000.
4. Ferrucci, P., Ferroresonance, *Cashier Tech.*, Schneider No. 190, ETC 190, March 1998.
5. Enrigth, W., Watson, N., and Nayak, O.B., Three phase five-limb unified magnetic equivalent circuit transformer models for PSCAD V3, *IPST '99 Proc.*, Budapest, 462–467, 1999.
6. Enrigth, W., Nayak, O.B., Irwin, G.D., and Arrillaga, J., An electromagnetic transient model of the multi-limb transformers using normalized core concept, *IPST'97 Proc.*, 93–108, 1997.
7. Finney, D., *Variable Frequency AC Motor Drive Systems*, Peter Peregrin Ltd. On behalf of the Institution of Electrical Engineers, IEE Power Engineering Series 8, London, U.K., 1988
8. McEachern, A., Power Quality Teaching Toy, Edition 2.0, 2005.
9. Hingorani, N.G. and Gyugyi, L., *Understanding FACTS: Concepts and Technology of Flexible AC Transmission Systems*, IEEE Press, New York, 1999.
10. Tolbert, L.M., Survey of harmonics measurements in electrical distribution systems, *IEEE IAS Annu. Meet.*, Oct. 6–10, 1996, San Diego, CA, 2333–2339.
11. Private communication from Hatch, Smart Predictive Line Controller, 2005.

3 Standardization of Harmonic Levels

3.1 INTRODUCTION

The most widespread standards for harmonic control worldwide are due to IEEE in the U.S. and IEC (International Electrotechnical Commission) in the European Union. In 1981, the IEEE issued Standard 519-1981,[1] which aimed to provide guidelines and recommended practices for commutation notching, voltage distortion, telephone influence, and flicker limits produced by power converters. The standard contended with cumulative effects but did little to consider the strong interaction between harmonic producers and power system operation.

The main focus of the revised IEEE-519 standard in 1992[2] was a more suitable stance in which limitations on customers regarding maximum amount of harmonic currents at the connection point with the power utility did not pose a threat for excessive voltage distortion. This revision also implied a commitment by power utilities to verify that any remedial measures taken by customers to reduce harmonic injection into the distribution system would reduce the voltage distortion to tolerable limits. The interrelation of these criteria shows that the harmonic problem is a system, and not a site, problem. Compliance with this standard requires verification of harmonic limits at the interface between utilities and customers, more commonly known as PCC.

Recommended total harmonic distortion (THD) levels for current and voltage signals were established in the 1992 revision of IEEE-519. Total and individual harmonic distortion levels were issued for customers on current and for utilities on voltage signals at the PCC. The TDD concept was created to better relate the THD to the demand current. The TDD is the total root sum square (RSS) of the harmonic current expressed in percentage of the nominal maximum demand load current. The standard also called for limiting commutation-notching levels at individual low-voltage customer locations. Communication interference with systems produced by low-voltage DC converters was addressed in the revised standard and IT limits for utilities were also established. All recommended IEEE-519-1992 limits were presented for different voltage levels encompassing 69 kV and below, 69.001 through 161 kV, and higher than 161 kV.

The 1992 edition of the IEEE harmonic standard thus advocates the joint contribution of utilities and customers to contend with harmonic emission and control matters. The compliance with recommended levels was deemed a convenient way to keep harmonic current penetration into the distribution system under control and permit the operation of equipment and devices that draw current in a nonlinear fashion. For over a decade, this standard has been the main reference for American utilities, customers, and manufacturers alike in trying to sustain the operation of the electrical systems within tolerable waveform distortion margins. Harmonic distortion limits were not specified for particular types of customers (industrial, commercial, or residential). The observance of these limits brought also the need for properly modeling utility systems and harmonic sources, which led to the development of expert software whose main characteristics are described in Chapter 8.

Similar to IEEE-519, IEC harmonic standards set limits at the utility–customer interface; they also set limits for customer equipment, in a clear reference to residential installations. After multiple revisions, the last IEC harmonic standard 61000-3-2[3] focused on limiting equipment consumption of harmonics. The equipment refers to single- and three-phase units with per-phase currents up to 16 A. Individual harmonic limits are required for every one of the four different classes of equipment considered — namely, A through D. Class D is regarded as a highly harmonic producer and its harmonic content is subject to a strict maximum harmonic current per unit of the current drawn at the main frequency. These are more stringent limits than for the other classes of equipment.

Regarding voltage distortion, compatibility[4] and planning[5] levels are specified for electrical networks to tie in with emission and immunity levels, respectively, in LV and MV installations. Compatibility levels are used as a reference for coordinating the emission and immunity of the equipment in LV and MV installations. Planning levels are used by system operators in evaluating the impact of all disturbing loads on the utility supply. For MV, compatibility levels are described in IEC 61000-2-12:2003-04.[4] Indicative values for planning levels along with definitions of LV, MV, HV, and EHV are given in IEC 61000-3-6:1996[5]: LV is 1 kV or less; MV is above 1 kV and below 35 kV; HV is from 35 to 230 kV; and EHV is above 230 kV.

Unlike IEEE-519, IEC considers the harmonic distortion assessment to cover short- and long-term effects.[5] The former are referred to as very short (3 s) events and the latter as short-period (10 min) events. Very short-time events are meant to account for disturbing effects on electronic devices that may be susceptible to harmonic levels lasting up to 3 s, excluding transients. Long-term effects account for thermal effects on equipment such as transformers, motors, cables, capacitor banks, etc. However, for statistical assessment, periods of 1 week or longer are recommended.[5-8]

Interharmonic (harmonic components not an integer of the fundamental frequency) voltage compatibility limits related to flicker in lighting devices are addressed in IEC 61000-2-2.[7] IEEE-519:1992 does not specifically set limits for interharmonics. However, it is expected that this issue will be included in the present revision of the standard, in accordance with Halpin.[10]

3.2 HARMONIC DISTORTION LIMITS

The rms value of a voltage waveform, considering the distortion produced by harmonic currents, is expressed by:

Standardization of Harmonic Levels

$$Vrms = \sqrt{\sum_{h=1}^{\infty} Vh^2} \qquad (3.1)$$

Likewise, the rms value of a sinusoidal current, taking into account the distortion created by the harmonic source currents, is given by:

$$Irms = \sqrt{\sum_{h=1}^{\infty} Ih^2} \qquad (3.2)$$

As defined in Chapter 1, total harmonic distortion is a parameter used in IEEE and IEC standards. For the sake of convenience, the definition of THD discussed in Chapter 1 is repeated here for voltage and current signals, respectively:

$$THD_V = \frac{\sqrt{\sum_{h=2}^{\infty} V_h^2}}{V_1} \qquad (3.3)$$

$$THD_I = \frac{\sqrt{\sum_{h=2}^{\infty} I_h^2}}{I_1} \qquad (3.4)$$

3.2.1 IN AGREEMENT WITH IEEE-519:1992

Per IEEE-519,[2] recommended harmonic distortion limits are to be verified through comparison with measurements at the PCC, i.e., the interface between the electric utility and the customer. Chapter 5 describes the relevant aspects involved in the measurements of harmonics. A significant issue is that levels can be exceeded by 50% under start-ups or unusual conditions with durations less than an hour. It additionally recommends the use of probability distribution functions from field measurements, stating that if limits are exceeded for only a "short" period, such a condition can be considered acceptable.

As portrayed in Figure 3.1, the recommended limits are a function of the system voltage level. For electric networks 69 kV and below, for example, the total voltage distortion is limited to 5%; no individual voltage harmonic should exceed 3%, as depicted in the illustration.

Concerning current harmonic distortion, IEEE-519 defines the limits as a function of the ratio between the short-circuit current at the PCC (I_{sc}) and the average current corresponding to the maximum demand during a period of 12 months (I_L). The recommended limits are summarized in Figure 3.2. Notice that

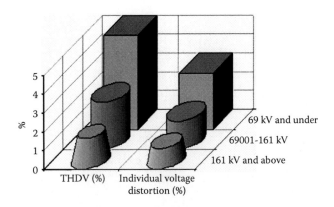

FIGURE 3.1 IEEE-519 voltage harmonic distortion limits.

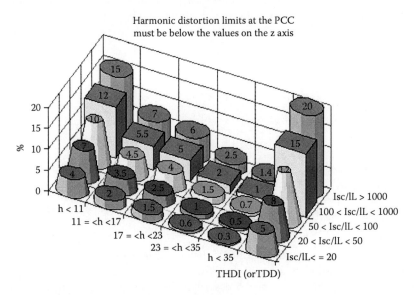

FIGURE 3.2 IEEE-519 current distortion limits.

the suggested limits become more stringent for decreasing I_{sc}/I_L ratios and increasing harmonic order.

The following aspects are to be noticed:

Regardless of the I_{sc}/I_L ratio at the PCC, all power generation equipment must meet the values given for an I_{sc}/I_L ratio < 20.

Even harmonics are limited to 25% of the odd harmonic limits.

Current distortions that result in a DC offset (e.g., half-wave converters) are not allowed.

Note that total harmonic distortion limits are expressed in terms of the total demand distortion or TDD, which refers to the electric demand during a period of 15 to 30 min.

Standardization of Harmonic Levels

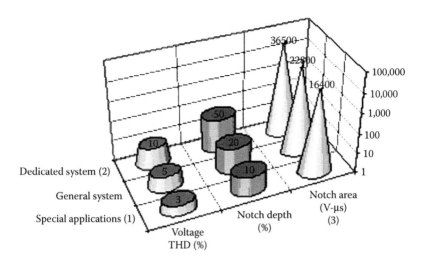

(1) Special applications include hospitals and airports
(2) A dedicated system is exclusively dedicated to converter loads
(3) In V-µs for rated voltage and current. If voltage is different than 480 V, multiply by V/480

FIGURE 3.3 Notching levels from IEEE-519. (Data from Standard 519-1992: "IEEE Guide for Harmonic Control and Reactive Compensation of Static Power Converters.")

The shown values are the maximum permissible limits under continuous operation. During start-up or unusual operating conditions lasting less than 1 hour, these limits can be exceeded by 50%.

In systems that use converters of more than six pulses, these limits can be increased by $\sqrt{\frac{q}{6}}$, where q is the number of pulses.

IEEE-519 also sets limits for commutation notching in LV power converters. Figure 3.3 depicts the allowed notch depth and notch area along with the corresponding THD.

IEEE-519:1992 practically establishes a joint involvement of the customer and the electric company to maintain harmonic distortion levels within limits that will enable the power distribution systems to operate within safe voltage distortion limits. Making customers play a part in this process assures the direct benefit of minimizing the negative effects that harmonic currents may have at their premises. Figure 3.4 outlines this relationship.

3.2.2 In Conformance with IEC Harmonic Distortion Limits

With regard to IEC, as described earlier, compatibility levels are used as a reference for coordinating the emission and immunity of the equipment in LV and MV installations. Planning levels are used by system operators in evaluating the impact of all disturbing loads on the utility supply. For the interface utility/customer short-

FIGURE 3.4 Relationship between customers and power utilities to achieve compliance with IEEE-519.

time (10 min) measurement results of voltage distortion, usually taken as the value related to the 95% probability weekly value,[5] must conform to planning levels. Planning levels are defined in IEC 61000-2-12;[4] higher level emissions reaching up to 11% for very short periods (3 s) are also considered. These are important in assessing very short-time effects of harmonics.

Figure 3.5 shows the IEC Compatibility[4] and Planning[5] individual voltage distortion levels for MV systems. As noted, THD levels for MV systems are somewhat higher than those from IEEE in Figure 3.1.

Regarding customer equipment, IEC standards specifically set harmonic levels aimed at protecting low-voltage systems at customer and utility installations. IEC

FIGURE 3.5 IEC voltage distortion limits.

Standardization of Harmonic Levels

61000-3-2[3] considers four different classes of equipment in establishing harmonic current limits:

Class A: balanced three-phase equipment; household appliances (excluding equipment) identified as class D; tools (except portable), dimmers for incandescent lamp (but not other lighting equipment), audio equipment; anything not otherwise classified

Class B: portable power tools

Class C: all lighting equipment except incandescent lamp dimmers

Class D: single phase, under 600 W, personal computer, PC monitor, TV receiver

Table 3.1 presents the limits for individual harmonic current for every one of the classified equipment classes. Limits are given in amperes for equipment classes A and B and in percentage of fundamental for class C. For class D, levels are specified in milliamperes per watt for equipment with a rated power exceeding 75 W but inferior to 600 W, or in amperes for equipment larger than 600 W. Notice that total harmonic distortion levels are not specified. For an application involving 230-V class D equipment subject to the maximum level of individual harmonic current, the total harmonic distortion would yield around 95%.[11]

Finally, IEC61000-2-2[7] defines compatibility levels for situations of interharmonic voltages occurring near the fundamental frequency. Specific types of loads are sensitive to the square of the voltage and exhibit a beat effect resulting in flicker. Figure 2 in reference 7 describes maximum interharmonic amplitudes (as a percentage of fundamental voltage) as a function of the difference between the interharmonic and the fundamental frequency (beat frequency). In essence, voltage oscillations between 1 and 4% are established for beat frequencies below 1 Hz or between 20 and 40 Hz (too small or too large differences) and voltage variations between about 0.2 and 1% are set for beat frequencies between 1 and 25 Hz for 120- and 230-V lamps.

Other effects of interharmonics and subharmonics, including their detrimental effects on underfrequency relays and harmonic torques in rotating machines that are not addressed in the IEC standards, are described in reference 8.

TABLE 3.1
IEC 61000-3-2 Harmonic Current Limits for Different Equipment Classes

Odd Harmonic n	Even Harmonic n	Max. Permissible Harmonic Current (A) Class A	Max. Permissible Harmonic Current (A) Class B	Max. Permissible Harmonic Current (% of Fundamental) Class C	Max. Permissible Harmonic Current (mA/W) 75 W < P < 600 W Class D	Max. Permissible Harmonic Current (A) P > 600 W
	2	1.08	1.62	2		
3		2.3	3.45	(30) × circuit power factor	3.4	2.3
	4	0.43	0.645			
5		1.14	1.71	10	1.9	1.14
	6	0.3	0.45			
7		0.77	1.155	7	1	0.77
	8 ≤ n ≤ 40	1.84/n	2.76/n			
9		0.4	0.6	5	0.5	0.4
11		0.33	0.495	3 (For all 11 ≤ n ≤ 39)	0.35	0.33
13		0.21	0.315		0.296	0.21
15 ≤ n ≤ 39		2.25/n	3.375/n		3.85/n	2.25/n

REFERENCES

1. Standard 519-1981: "IEEE Guide for Harmonic Control and Reactive Compensation of Static Power Converters."
2. IEEE Std. 519-1992: Recommended Practices and Requirements for Harmonic Control in Electric Power Systems.
3. IEC 61000-3-2 (2001-10): Electromagnetic compatibility (EMC) — part 3-2: limits — limits for harmonic current emissions (equipment input current $\leq$ 16 A per phase).
4. IEC 61000-2-12, 2003-04: Electromagnetic compatibility (EMC) — part 2-12: compatibility levels for low-frequency conducted disturbances and signaling in public medium-voltage power supply systems.
5. IEC 61000-3-61996: Assessment of emission limits for distorting loads in MV and HV power systems, technical report type 3.
6. IEC 61000-4-30, 2003: Power quality measurement methods.
7. IEC 61000-2-2: Electromagnetic compatibility — part 2-2, environment compatibility levels for low-frequency conducted disturbances and signaling in public and low-voltage power supply systems, 2002.
8. Joint WG CIGRE C4.07/CIRED, Power quality indices and objectives, final WG Report, January 2004, Rev. March 2004.
9. Fuchs, E.F., Roesler, D.J., and Masoum, M.A.S., Are harmonic recommendations according to IEEE and to IEC too restrictive? *IEEE Trans. Power Delivery*, 19(4), Oct. 2004.
10. Halpin, M., Harmonic modeling and simulation requirements for the revised IEEE Standard 519-1992, *2003 IEEE Power Eng. Soc. General Meet. Conf. Proc.*, July 13–17, 203, Toronto, Ontario, Canada.
11. Ward, J. and Ward, D., Single phase harmonics, PSER EMI, Power Quality, and Safety Workshop, April 18–19, 2002.

4 Effects of Harmonics on Distribution Systems

4.1 INTRODUCTION

By the turn of the century, electronic loads must have hit a remarkable 50% participation in the U.S. electrical demand, with much of that growth involving the residential sector, as envisaged by De Almeida.[1] This is a clear indication that residential customers are joining industrial and commercial customers as harmonic current generators at a rapid pace. Considering the limited awareness of residential customers on harmonics created by household equipment (multiple TV sets, computers and entertaining devices, fluorescent lighting, etc.), power utilities may find it increasingly difficulty to set up rules for implementing remedial actions at this user level. Even in commerce and industry, the concept of harmonic filters is far from adequately well known. This is often due to the lack of information on the effect that harmonics producing nonlinear loads can impose on sensitive industrial processes and equipment and commercial applications.

As a result, a considerable number of electricity users are left exposed to the effects of harmonic distortion on industrial, commercial, and residential loads. In a broad manner, these can be described as the following.

4.2 THERMAL EFFECTS ON TRANSFORMERS

Modern industrial and commercial networks are increasingly influenced by significant amounts of harmonic currents produced by a variety of nonlinear loads like variable speed drives, electric and induction furnaces, and fluorescent lighting. Add to the list uninterruptible power supplies and massive numbers of home entertaining devices including personal computers.

All of these currents are sourced through service transformers. A particular aspect of transformers is that, under saturation conditions, they become a source of harmonics. Delta–wye- or delta–delta-connected transformers trap zero sequence currents that would otherwise overheat neutral conductors. The circulating currents in the delta increase the rms value of the current and produce additional heat. This is an important aspect to watch. Currents measured on the high-voltage side of a delta-connected transformer will not reflect the zero sequence currents but their effect in producing heat losses is there.

In general, harmonics losses occur from increased heat dissipation in the windings and skin effect; both are a function of the square of the rms current, as well as from eddy currents and core losses. This extra heat can have a significant impact in reducing the operating life of the transformer insulation. Transformers are a particular case of power equipment that has experienced an evolution that allows them to operate in electrical environments with considerable harmonic distortion. This is a K-type transformer. Because losses and K-type transformers are further described in Chapter 9, here we only stress the importance of harmonic currents in preventing conventional transformer designs from operating at rated power under particular harmonic environments. In industry applications in which transformers are primarily loaded with nonlinear loads, continuous operation at or above rated power can impose a high operating temperature, which can have a significant impact on their lifetime.

4.2.1 Neutral Conductor Overloading

In single-phase circuits, return currents carrying significant amounts of harmonic components flow through transformer neutral connections increasing the rms current. Furthermore, zero sequence currents (odd integer multiples of 3) add in phase in the neutral. Therefore, the operation of transformers in harmonic environments demands that neutral currents be evaluated in grounded-wye connected transformers to avoid the possibility of missing the grounding connection as a consequence of overloading. In balanced three-phase, four-wire systems, there is no current on the neutral, for which the presence of neutral currents under these conditions should be attributed to the circulation of zero sequence harmonics, which are mostly produced by single-phase power supplies.

In systems that are not entirely balanced, the unbalanced current circulates on the return (neutral) conductor. Because this conductor is usually sized the same as the phase conductors for being able to handle unbalanced currents comfortably, it may experience overheating if those currents are subsequently amplified by zero sequence currents. Large numbers of computers in office buildings make a formidable source of harmonic currents produced by their electronic switched power supplies.

A common practice is to size neutral conductors to carry as much as two times the RMS current that phase conductors can take. Monitoring temperature increase on the neutral conductor of transformers might be a good start to detect whether zero sequence harmonic currents are not overstressing neutral connections. This is true as long as the system does not incur increased levels of current unbalance that would produce a temperature rise in neutral conductor temperature.

4.3 MISCELLANEOUS EFFECTS ON CAPACITOR BANKS

4.3.1 Overstressing

Increased voltage can overstress and shorten the life of capacitor banks. Voltage, temperature, and current stresses are the drivers of capacitor bank conditions that

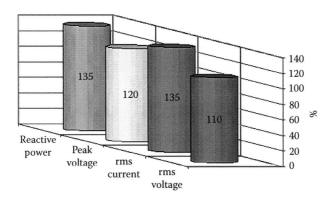

FIGURE 4.1 Allowed operation limits for shunt capacitor banks. (Data from IEEE 18-2002, IEEE Standard for Shunt Power Capacitors, Institute of Electrical and Electronics Engineers, October, 2002.)

lead to dielectric breakdown. The output reactive power from a capacitor bank varies with the square of the voltage, as described by:

$$VAR = \frac{V^2}{X_c} \qquad (4.1)$$

Operating voltage can increase in distribution systems under light load conditions or when fuse links operate to isolate a failed capacitor unit, leaving the remaining units exposed to an overvoltage condition. For example, a 5% increase in the nominal voltage of a capacitor unit would cause it to deliver $(1.05)^2 = 1.1$ or 110% of rated reactive power. Harmonic distortion is definitively another factor that contributes to impose voltage stresses on capacitor banks. This is a serious condition in industrial facilities with unfiltered large power converters.

IEEE-18[2] allows for the increase of fundamental operating parameters in capacitor banks as illustrated in Figure 4.1. The y-axis indicates values in percentage of nominals. These operating limits are for continuous operation. Thus, it will be important to take into account these limits also in the design of harmonic filters because capacitor banks in single–tuned filters are meant to act as a sink for the entire amount of harmonic currents of the corresponding tuned frequency.

4.3.2 Resonant Conditions

As further addressed in Chapter 6, resonant conditions involve the reactance of a capacitor bank that at some point in frequency equals the inductive reactance of the distribution system, which has an opposite polarity. These two elements combine to produce series or parallel resonance. In the case of series resonance, the total impedance at the resonance frequency is reduced exclusively to the resistive circuit component. If this component is small, large values of current at such frequency will be developed. In the case of parallel resonance, the total impedance at the resonant frequency is very large (theoretically tending to infinite). This condition may produce a large overvoltage between the parallel-connected elements, even under small harmonic currents. Therefore, resonant conditions may represent a hazard for solid insulation in cables and transformer windings and for the capacitor bank and their protective devices as well.

Resonant frequencies can be anticipated if the short-circuit current level at the point where the capacitor bank is installed is known, following Equation (4.2):

$$h_r = \sqrt{\frac{kVA_{short\_circuit}}{kVAR_{cap\_bank}}} \qquad (4.2)$$

where
h_r is the resonant frequency as a multiple of the fundamental frequency.
$kVA_{short\_circuit}$ is the short-circuit power available at the site.
$kVAR_{cap\_bank}$ is the reactive power rating of the capacitor bank.

Note how changing any of these parameters can shift the resonant frequency. This is a practice actually used sometimes in certain applications involving excessive heating in transformers connected to non-linear loads.

If this frequency coincides with a characteristic harmonic present at the site, that current will see a large upstream impedance and the existing voltage harmonic distortion will be amplified. Balda et al.[3] suggested that capacitor banks can be applied without concern for resonance conditions as long as the nonlinear load and capacitor bank are less than 30 and 20%, respectively, the rated kilovoltamperes of the transformer, assuming a typical transformer impedance around 5 to 6%. Otherwise, the capacitors should be used as a harmonic filter, with a series reactor that tunes them to one of the characteristic harmonics of the load. Generally, fifth and seventh harmonics are the most commonly found and account for the largest harmonic currents.

4.3.3 Unexpected Fuse Operation

As mentioned earlier, rms voltage and current values may increase under harmonic distortion. This can produce undesired operation of fuses in capacitor banks or in laterals feeding industrial facilities that operate large nonlinear loads. Capacitor banks can be further stressed under the operation of a fuse on one of the phases, which leaves the remaining units connected across the other phases. They are thus left subject to an unbalanced voltage condition that can produce overvoltages and detune passive harmonic filters if they are not provided with an unbalance detection feature.

An important aspect to look after with the advent of cogeneration and microturbine schemes using power inverters with electronic switching technology will be their harmonic current contribution and how this may affect the operation performance of islanding protective relays.

4.4 ABNORMAL OPERATION OF ELECTRONIC RELAYS

VFD operation leading to shut-down conditions is often experienced in applications involving oil fields in which solid material (sand) abruptly demand higher thrust power, mining works in which sudden increases in lifting power occur, and high inertia loads, among others. In all these cases, the protective relays trip as a response to overcurrents exceeding the established settings. Similar effects can be experienced under the swift appearance of harmonic distortion on current or voltage waveforms exceeding peak or rms preset thresholds. Therefore, when protective relays trigger during the operation of a nonlinear load, harmonic distortion should be assessed. It might well be that an unpredicted overloading condition is the cause of the unexpected operation, but often increased harmonic levels following nonlinear load growth are the reason for similar behavior.

On the other hand, third harmonic currents produced by severe line current unbalance may cause nuisance relay tripping in VFD applications. Therefore, nuisance and missed relay tripping in installations with nonlinear loads should be assessed by checking the harmonic distortion levels and by inspecting the relays for possible threshold-setting fine-tuning. The onset of this type of occurrence in industrial installations may be used as an warning to start considering harmonic filtering actions.

4.5 LIGHTING DEVICES

Chapter 2 presented some examples of harmonic current generation in fluorescent lamps using magnetic and electronic ballast. This phenomenon, though, does not produce a self-impact on lighting luminosity levels. It appears that frequency components that are a noninteger multiple of the fundamental frequency, also called interharmonics, are more prone to excite voltage oscillations that lead to light flickering. The main sources of interharmonics are the cycloconverters widely used in the steel, cement, and mining industries, as well as arc welders and furnaces. According to the joint IEEE TF and the CIGRE/CIRED WG on Interharmonics,[4] cycloconverters have characteristic frequencies of

$$f_i = (p1 \cdot m \pm 1) f_1 \pm p2 \cdot n \cdot f_o \quad (2) \tag{4.3}$$

where f_i is the interharmonic frequency.
where f_i is the fundamental frequency.
$p1$ and $p2$ are number of pulses on the rectifier and output sections, respectively.
m and n are integers 0, 1, 2, 3, ... but not 0 at the same time.
f_o is output frequency of the cycloconverter.

Light flicker is one of the main impacts of interharmonics due to the modulated steady-state interharmonic voltage on the power frequency voltage. According to reference 4, the rms voltage fluctuations that can be produced by interharmonic phenomena can be expressed by:

$$U = \sqrt{\frac{1}{T}\int_0^T [\sin 2\pi f_1 t) + a\sin(2\pi f_i t)^2]dt} \qquad (4.4)$$

where a is the amplitude of the interharmonic voltage in p.u. with $a = 1$ for the fundamental frequency.

It should be noted that incandescent lamps are more sensitive to flickering responding to rms voltage variations and fluorescent lamps are more sensitive to peak voltage fluctuations.

4.6 TELEPHONE INTERFERENCE

The common construction of telephone lines built underneath power conductors on electric utility distribution poles makes them prone to a number of interference phenomena. Arrillaga et al.[5] describe the inductive, capacitive, and conductive interference that can take place between a power and a telephone line. In Chapter 1, the telephone influence factor (TIF) and the IT product were described as some of the power quality indices used by IEEE-519[6] to address and recommend limits on the telephone interference issue under harmonic distortion. According to reference 3, an IT product over 25,000 will probably cause interference problems.

4.7 THERMAL EFFECTS ON ROTATING MACHINES

Similar to transformers, rotating machines are exposed to thermal effects from harmonics. Because the effective resistance of a conductor goes up as frequency rises, a current wave rich in harmonics may cause greater heating on winding conductors than a sine wave of the same rms value. The overall effect can lead to a decreased transformer lifetime. The most significant aspects of rotating machine losses due to harmonics are described in Chapter 9.

4.8 PULSATING TORQUES IN ROTATING MACHINES

Additionally, magnetomotive forces (mmf) induced by positive and negative sequence harmonics interact with the nominal frequency mmf force creating torque components of different frequencies (as described by Escobar and De la Rosa[7]). This may lead to problems on the shaft of rotating machines subject to the influence of harmonic torsional pairs including:

Equipment fatigue
Unexplained operation of "mechanical fuses" (bolts used to bond together turbine and generator shafts)
Increased vibration
Bearing wear out

Effects of Harmonics on Distribution Systems

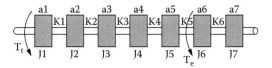

FIGURE 4.2 Model of a mechanical turbine-generator system.

As an illustration, we analyze the case of a turbine-generator set whose mechanical shaft model is illustrated in Figure 4.2.[7]

The equation of motion of a mechanical system describing the developed torque as a function of angular displacement,, and inertia constant, J, can be described per:

$$J\ddot{\theta} + D\dot{\theta} + K\theta = T \tag{4.5}$$

By disregarding the damping matrix, through modal analysis we can describe the eigenvalues (natural frequencies of the system) as follows:

$$\Delta = \left| K - J\omega^2 I \right| = 0 \tag{4.6}$$

$$\lambda = \left| \omega_1^2 \omega_2^2 \omega_3^2 \quad \omega_i^2 \right| \tag{4.7}$$

Assuming that matrix D is composed of viscous damping and that it has the form of a linear combination of matrices J and K, we can write:

$$\theta = Xq \tag{4.8}$$

where the eigenmatrix X includes eigenvectors of every mode and q is the vector of the new variables.

By using orthogonality properties of eigenvectors, we can uncouple the equation of motion:

$$J^{-1}(X^t J X \ddot{q} + X^t D X \dot{q} + X^t K X q = X^t T) \tag{4.9}$$

which leads to n decoupled second-order differential equations of the form:

$$\ddot{q} + 2\zeta_i \omega_i \dot{q} + \omega_i^2 q = Q_i \tag{4.10}$$

We integrate this equation to get the simulation in the time domain.

Let us now look at the steady-state amplitude of the electric torque polluted with one oscillating component. As input, we have one pulsating component of electrical torque as:

$$T_e = T_b + T_h \sin(\omega_f t) \qquad (4.11)$$

where ω_f is the frequency of the pulsating torque. Then Q_i has the form:

$$Q_i = C_{1i} + C_{2i} \sin(\omega_f t) \qquad (4.12)$$

The solution for a steady-state sinusoidal excitation torque (Q_i) for the uncoupled differential equation is:

$$q_i(t) = \frac{C_{1i}}{\omega_i^2} + |q|\sin(\omega_f t - \varphi) \qquad (4.13)$$

where phasor $|q|$ is:

$$q = \frac{C_{21}}{\omega_i^2 \sqrt{\left(1 - \frac{\omega_f^2}{\omega_i^2}\right)^2 + \left(2\zeta_i \frac{\omega_f}{\omega_i}\right)^2}} \qquad (4.14)$$

As an example, let us assume a single turbine–generator set subject to a harmonic load around 16% the capacity of the generator. This load is connected to the generator bus through a 5-MVA transformer, as illustrated in Figure 4.3, where the assumed natural frequencies are also indicated.

FIGURE 4.3 Turbine-generator set example.

TABLE 4.1
Current and Electrical Torque Spectra for Branch Bus 1 to Bus 2

Harmonic Order	Current Amplitude (%)	Electric Torque Amplitude (%)
1	100	100
5	2.247	
6		**1.162**
7	1.492	
11	0.89	
12		**0.8137**
13	0.742	
17	0.558	
18		**0.524**
19	0.497	
23	0.408	
24		**0.390**
25	0.375	
29	0.322	
30		**0.313**

Using the preceding analytical approach, we calculate the electric torque for the generator using the harmonic spectrum of the VFD converter and obtain the results presented in Table 4.1 and in Figure 4.4, in which the electric torque is plotted together with the harmonic spectrum of the VFD converter. Note the occurrence of electric torques at the intermediate frequencies of harmonic pairs 5–7, 11–13, 17–19, etc., which are characteristic of the six-pulse converter and are shown in bold face type in Table 4.1. The elevated torques at these frequencies can be regarded as mechanical resonant spectral components when coinciding with the assumed natural frequencies of the system indicated in Figure 4.3. This situation may lead to a severe increase in vibration amplitudes.

FIGURE 4.4 Harmonic spectrum showing calculated electric torque amplitudes for the described example.

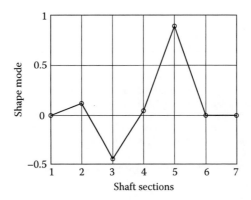

FIGURE 4.5 Calculated vibration mode shapes for the 12th harmonic natural oscillation frequency.

TABLE 4.2
Calculated Torques for the Different Shaft Sections

Shaft Section	Shaft Torque in p.u.
1–2	0.019
2–3	0.411
3–4	0.145
4–5	0.14
5–6	0.031
6–7	1.4×10^{-5}

It is important to state that the calculations presented involved a THD at the synchronous generator terminals of around 3.2%, well below the 5% permitted by standard IEEE-519[6] and the nonlinear load was less than 50% of the rated generator power.

Taking only the eigenvector corresponding to the natural frequency around the 12th-order harmonic, the vibration mode shapes depicted in Figure 4.5 are calculated.

From the shape mode deflections, we can anticipate major oscillations between elements 2 and 3 due to their different polarity. This means that torques act in opposite directions. The lowest value of shaft torque can be expected in elements 6 and 7, where there is almost no mode shape deflection between these two elements. Shaft torques will be expected in all other shaft sections. Using Equation (4.15), the shaft torques, T_m, for all the shaft sections are calculated and presented in Table 4.2.

$$T_m = K_{(i,i+1)}(\theta_{i+1} - \theta_i) \tag{4.15}$$

Figure 4.6 shows the calculated shaft torque in the different shaft sections.

Effects of Harmonics on Distribution Systems

(a) Shaft torque in section 1-2

(b) Shaft torque in section 2-3

(c) Shaft torque in section 3-4

FIGURE 4.6 Calculated shaft torques for the example of Figure 4.2. *Continued.*

(d) Shaft torque in section 4-5

(e) Shaft torque in section 5-6

(f) Shaft torque in section 6-7

FIGURE 4.6 *Continued.*

The mechanical damping is usually low and depends on mechanical design and operative condition. Thus, from the previous results it appears that a possible way to avoid a mechanical resonance could be achieved by modifying the natural frequencies of the mechanical system — for example, by diminishing the inertia (J) and spring (K) constants to reduce residual vibrations further. This would require a joint and open participation of manufacturers and industry to define practical scenarios and take appropriate actions.

4.9 ABNORMAL OPERATION OF SOLID-STATE DEVICES

Under unexpected circumstances, harmonic distortion can also lead to multiple zero crossing of the voltage waveform disturbing the operation of control systems that use the zero crossing as a timing or counter reference. Additional problems may include interference on motor speed controllers and abnormal VFD operation when rms voltage or current values are altered by harmonic distortion. This often leads to the need to apply remedial measures at the source side of the converters and/or readjustment of the protection threshold settings.

4.10 CONSIDERATIONS FOR CABLES AND EQUIPMENT OPERATING IN HARMONIC ENVIRONMENTS

In light of a steady increase in harmonic distortion in power systems, the specifications and selection criteria of electrical equipment in industrial installations should be revisited. For example, when distributed generators operate in facilities in which nonlinear loads predominate, the response of conventional generator designs with large subtransient reactances has been shown to be ineffective and often ends in equipment failure after exposure to severe thermal stresses. There must be a threshold for operating parameters that allows equipment to withstand worst-case operating conditions regarding harmonic content and its diverse effects. This goes beyond the theoretical steady-state operation mostly assumed when specifying equipment and network components.

4.10.1 GENERATORS

Generators used in the electric power industry are fundamentally designed to feed linear loads. However, when the type of load is predominantly nonlinear, generation systems must comply with certain requirements that allow them to operate in stable conditions and without being exposed to excessive heating and torsional torque vibrations, which can make them exceed their permissible operating limits.

Essentially, a nonlinear load produces a voltage waveform distortion at the generator terminals; this imposes the following consequences in the operation of a generator:

- Production of positive and negative sequence current contributions that generate torsional torques and vibration mode shapes on the motor axis. The thermodynamic forces created in the rotor can prematurely wear out shaft bearings.
- Voltage waveform distortion on the supply circuit to the excitation system; this can produce voltage regulation problems.
- Excessive negative sequence currents; these can contribute to increased voltage unbalance.

As a reference, the following list describes some of the characteristics of synchronous generators found to perform adequately in land and offshore oil well ESP applications. This relates to cases where VFDs make the largest portion of the load and where isolated DGs source all or most of the demanded power:

Independent static excitation system is present.
Rotor is furnished with ammortisseur (damper) copper bars.
Transient reactance, X_d', is between 16 and 18%.
Direct axis subtransient reactance X_d'', is between 13 and 15%. Notice that these values are considerably lower than typical synchronous generators for linear load applications, which are designed with subtransient impedances around 25%.
Insulation (rotor and stator) is according to American Petroleum Standard-546.
Power factor 0.85 is lagging. Here it is important to stress that industrial systems that involve large VFDs or significant numbers of small units may run at power factors close to unity and in some cases even on leading power factors.
Operation with nonlinear loads is satisfactory. This means that a generator must withstand a maximum THD of 8% across its terminals. Notice that this THDV is in excess of the 5% recommended by IEEE for general distribution systems, 120-69000 V, as discussed in Chapter 3.
An open-circuit voltage shows waveform distortion inferior to 2%.
Generators are provided with an oversized damper winding consisting of copper bars to properly handle the additional heating caused by harmonic currents.
Static brushless type excitation systems with (around 10%) oversized rectifying diodes are present.
PMG (permanent magnet generator) types of excitation systems are present.
The regulator must be able to handle harmonic distortion typical of 12-pulse types of converters.

This list reveals features of synchronous generator designs that industry should consider to assure satisfactory performance of generating units in electrical environments highly exposed to harmonics.

4.10.2 Conductors

Power conductors used in distribution systems must be able to carry fundamental and harmonic currents without developing conductor overheating that would be translated to excessive losses. For this, it is important to select conductor sizes considering a permanent steady-state condition over current factor of at least 125%, following National Electrical Code NEC-1996, articles 430-24 and 220-10(b). In the latter article, it is recommended that conductor sizes be chosen to withstand 125% of the continuous currents plus the nonpermanent ones.

Also, in installations with shielded cables where shielding is grounded at intervals, it is important to consider a margin to account for the effect of induced currents in the power conductors. An additional 10% to the current specified in the former paragraph holds reasonable. However, oversizing conductors to take additional currents up to 100% of rated values in some VFD applications is sometimes common in industrial networks. This may occur in cases when the electric networks are designed for accommodating future load expansions.

For networks operating in harmonic environments, use of specialized software to determine conductor ampacity in the presence of waveform distortion under worst-case scenarios is highly recommended. These results can also be used for protection coordination purposes.

4.10.3 Energy-Metering Equipment

The impact that harmonic distortion can have on induction disk meters is an area of present debate in the technical community. The question that technical working groups are trying to resolve is not simply whether the induction disk in a watt-hour meter runs faster or slower because these conditions may change depending on the magnitude and order of existing harmonic and DC offset on voltage and current signals at the metering point. Efforts are focused on trying to define what actions can allow the improvement of active and reactive power metering under severe harmonic distortion.

Fortunately, utilities show increasing concern for timely detection of excessive harmonic distortion situations and seeking compliance with recommended limits through the application of corrective actions. The application of harmonic mitigation measures at specific locations where nonlinear loads exist will actually prevent extreme voltage distortion from penetrating the network and reaching metering points at adjacent facilities.

REFERENCES

1. De Almeida, A., Understanding power quality, *Home Energy Magazine* Online, Nov./Dec. 1993, http://homeenergy.org/archive/hem.dis.anl.gov/eehem/93/931113.html
2. IEEE 18-2002, IEEE Standard for Shunt Power Capacitors, Institute of Electrical and Electronics Engineers, Oct. 2002.

3. Wagner, V.E., Balda, J.C., Griffitki, D.C., McEachern, A., Barnes, T.M., Hartmann, D.P., Phileggi, D.J., Emannuel, A.E., Horton, W.F., and Reid, W.E., Effects of harmonics on equipment, *IEEE Trans. Power Delivery*, 8(2), April 1993, Report of the IEEE Task Force on the Effects of Harmonics on Equipment.
4. IEEE Interharmonic Task Force, Cigré 36.05/CIRED 2 CC02 Voltage Quality Working Group, Interharmonics in Power Systems, Dec. 1997.
5. Arrillaga, J., Bradley, D.A., and Bodger, P.S., *Power System Harmonics*, John Wiley & Sons, New York, 1985.
6. IEEE Std. 519-1992, Recommended Practices and Requirements for Harmonic Control in Electric Power Systems.
7. Escobar, J.C. and De la Rosa, F., Shaft torsional vibrations due to nonlinear loads in low-capacity turbine units, *2001 IEEE Power Engineering Society Meeting Proceedings,* Vancouver, B.C., July 15–19, 2001.

5 Harmonics Measurements

5.1 INTRODUCTION

The entire issue of power system measurements is aimed at collecting relevant data for assisting utility planning and operation in a number of aspects key to the efficient transmission and distribution of electric energy. It is also intended to provide reliable energy consumption metering at industrial, commercial, and residential facilities.

A glimpse into the operation of a power system can allow us to realize the many instances when measurements are required. For example, the energy trading among different interconnected utilities/cooperatives requires reliable power delivery measurements that quantify the number of energy blocks that are bought and sold. The substation engineer looks at multiple panel instrumentation to guarantee that voltage and frequency are kept within specified limits and that the current on the different feeders follows the predicted demand, which must match the capacity of the substation transformer banks. Power factor is also observed to assure a proper balance between active and reactive power to minimize losses in the distribution system.

As loads fluctuate during the day in response to different demand patterns, utilities switch capacitor banks on and off to keep the voltage profile within tolerable limits. Under light load conditions, there is no need for reactive power compensation; this typically occurs during nighttime. As the load picks up, so does the voltage drop along distribution feeders and, at some distance from the substation, voltage may tend to decrease below permissible limits. It is then when strategically placed capacitor banks or inductive voltage regulators are "switched on" to help raise the voltage profile back to nominal values. The active and reactive power measurements at the substation are therefore key to energy dispatch operators to keep voltage regulation within tight limits.

Another relevant measurement aspect is protection device coordination, which follows pre-established settings that allow protective devices to open as a response to large currents identified as faults. However, some of them, namely, distance relays, have the capability to carry out current measurements and, using the information from the fault current, they can provide an approximate distance to the fault. SCADA (supervisory control and data acquisition) systems communicate with substation and feeder remote terminal units, smart relays, and substation automation systems to monitor real-time status of the network and provide remote control of devices such as switches, capacitor banks, and voltage regulators. The list involving measurement and monitoring of electric parameters can go on and on.

With the steady increase of solid-state electronics in industrial, commercial, and residential facilities, utilities face an increasingly challenging task to carry out reliable measurements due to the waveform distortion on voltage and current signals. This unfair race between fast growing customers joining the crowd of harmonic generators and utilities struggling to adapt appropriate measuring schemes can soon leave utilities far behind. Decisive efforts to control harmonic currents within industry limits before they converge at and disturb distribution substation monitoring equipment must be undertaken.

At harmonic source locations, the problem can be even worse. The unfiltered higher–frequency components of current at harmonic-producing loads may not give rise only to measurement equipment inaccuracies when they reach or exceed certain thresholds.[1,2] They can produce communication interference,[3] equipment heating problems, false protective device tripping, and even instability conditions on voltage regulation systems in synchronous generators. This is particularly true in installations where customer substation transformers are loaded with mostly nonlinear loads.

An even more delicate problem arises when a customer generates significant harmonic waveform distortion that affects adjacent utility customers. Because all customers can be regarded as harmonic producers to some extent, utilities may find it difficult to pinpoint the right location of the harmonic source, particularly when resonant networks come into play. Utilities may need to carry out measurements at a number of locations involving suspected customers before they can decide the source of the problem to start discussing remedial measures.

Aspects that require a careful standpoint are those related to adequate measurement periods, transducers, and correct measuring equipment. This chapter provides a general discussion on the most significant aspects to observe regarding harmonic measurements.

5.2 RELEVANT HARMONIC MEASUREMENT QUESTIONS

5.2.1 WHY MEASURE WAVEFORM DISTORTION

From the electric utility perspective, the general objectives for conducting harmonic measurements may be summarized as follows:

 To verify the order and magnitude of harmonic currents at the substation and at remote locations where customer harmonic sources may be affecting neighboring installations
 To determine the resultant waveform distortion expressed in the form of spectral analysis
 To compare the preceding parameters with recommended limits or planning levels
 To assess the possibility of network resonance that may increase harmonic distortion levels, particularly at or near capacitor banks
 To gather the necessary information to provide guidance to customers in controlling harmonic levels within acceptable limits

Harmonics Measurements

To verify efficacy of implemented harmonic filters or other corrective schemes
To determine tendencies in the voltage and current distortion levels in daily, weekly, monthly, etc., basis

5.2.2 How to Carry out Measurements

As mentioned in Chapter 3, IEEE standard 519-1992[3] provides a general description of key features to take into account when conducting harmonic measurements. It does not, however, underline other aspects regarding duration, integration times, and statistical treatment of collected data. In its current distortion limits section, IEEE 519 suggests that the 15- (or 30-) min maximum demand averaged over a 12-month period should be used as the load current, I_L, to determine the ratio, Isc/I_L. This important aspect warns utilities about the need for keeping monthly records of maximum demand to assess total demand distortion properly.

IEC61000-4-7, edition 2,[4] considers the measurements of voltage and current to obtain spectral content up to the 40th harmonic using 200-ms measurement windows. The preferred test instrument must be based upon rms calculation of each performance index over a synchronous contiguous 12-cycle window. The 12-cycle window has been adopted in the IEC standards for 60-Hz systems. This 12-cycle data can then be processed into 3-s, 10-min, and 2-h interval data for each index. Note that rms index values would tend to decrease if larger measurement intervals were used.

Measurement windows are grouped and smoothed using a 1.5-s first-order filter, whose value (for each individual harmonic group) is compared against the limits established in the four test classes (A through D) of IEC 61000-3-6.[5,6] Power measurement is included in the setup because it is the basis for limit calculations for class D equipment. Thus, the measurement equipment is rather sophisticated because it must meet stringent design requirements.

Compatibility levels presented in Chapter 3 refer to harmonic levels sustained for periods up to 3 s to account for interference on sensitive electronic devices and up to 10 min to account for thermal effects on miscellaneous equipment and cables. Therefore, harmonic measurements must consider these needs to seek compliance with compatibility levels.[7]

5.2.3 What Is Important to Measure

If a utility engineer needed to decide the parameters to consider in evaluating harmonic distortion problems, most likely the decision would involve voltage and current waveforms. This is indeed the right choice because other parameters, such as real, reactive, and total power; energy; and even unbalance, can be calculated from these two quantities. As discussed in Chapter 1, distorted voltage and current waveforms can be expressed as Fourier or other time series. Harmonic distortion and all power quality indices described in Section 1.5 can in fact be determined from these two basic parameters. Nevertheless, power quality monitoring equipment is presently designed to directly provide peak and true rms voltage, current, and power quantities along with harmonic indices comprising total and individual harmonic distortion and transformer K factor, among others.

Interharmonics, noninteger spectral components, and subharmonics, spectral components with frequencies below the fundamental power frequency as described in Chapter 2, are not easy to characterize. IEC Standard 61000-4-7[4] provides definitions and signal processing recommendations for harmonic and interharmonic measurements.

5.2.4 Where Should Harmonic Measurements Be Conducted

Harmonic distortion occurrence in an electrical installation can sometimes be assessed through a simple inspection of the types of loads at a given customer installation. All this requires is familiarity with the characteristic harmonic spectrum of each type of common nonlinear load, as described in Chapter 2. However, considering additional waveform distortion caused by transformer saturation or resonant conditions, a more precise evaluation should be carried out. This involves direct measurements at selected locations — for example, the point of common coupling (PCC) described in Chapter 3 and the node where nonlinear loads are connected.

It is understandable that the main location where measurements are to be conducted is the customer–utility interface. This is so because compliance with IEEE and IEC harmonic limits must be verified at this location. In customer-owned transformer locations, the PCC is the point where the utility will meter the customer, generally the high-voltage side of the transformer. If the utility meters the low-voltage side, then this becomes the PCC.

Also, measurements at LV-connected equipment locations are required when compliance with IEC 61000-3-2[8] (which covers all electrical and electronic equipment with an input current up to 16 A per phase) is sought or when harmonic filtering schemes must be designed at nonlinear loads locations. This is more likely to occur in the industrial or commercial environment where large harmonic-producing loads are operated and served from transformers feeding other sensitive loads. Other instances in which harmonic measurements would be required are when studies are conducted to determine the reasons for abnormal operation or premature failure of equipment, unexpected relay protection tripping, or excessive telephone interference.

5.2.5 How Long Should Measurements Last

The decision on the optimal period to conduct harmonic measurements may appear somewhat complicated. The reasons for this are diverse. In residential circuits, due to similarity in the types of electronic loads, the expected spectral content may be easily characterized in short-term measurements. However, care must be exercised when the feeder that supplies residential customers is the same from where large commercial/industrial installations are served. If commercial installations are involved, it may be possible to anticipate the types of harmonics because they will typically be linked to fluorescent lighting and power sources from diverse LV electronic equipment.

Harmonics Measurements

Industrial installations, however, are a special case because they are usually composed of a mix of loads having a diversity of spectral contents, which may require long-term measurements to characterize harmonic content. This need may become more obvious if cyclic loads exist because measurements to characterize harmonics at the PCC would need to encompass all, or at least the most significant, duty cycles. Long-term measurements may also be required when investigating or trying to resolve the origin of suspicious disturbances affecting a number of customers.

IEEE 519 guidelines do not specify a definite measurement period for capturing harmonic waveform distortion. Under steady-state operation and where no loading variations occur, a few minutes recording may be sufficient and averaging over a few seconds should meet the requirements. However, due to the changing nature of loads in most situations, measurements over a few days may be needed to assure that load variation patterns and their effects on harmonic distortion are considered.

IEC 61000-2-2[9] suggests assessment periods of 1 week for 10-min values (Uh, sh related to long-term effects) and one day for 3-s values (Uh, vs related to very short-term effects) and to allow taking into account daily work shift patterns and participation of different types of loads in the data collection. Long-term effects relate to thermal effects on different kinds of equipment such as transformers, motors, capacitor banks, and cables from harmonic levels sustained for at least 10 min. Very short-term effects relate to disturbing effects on vulnerable electronic equipment by events lasting less than 3 s, not including transients. Statistical handling of data is carried out in the form of 95 or 99 percentile of daily or weekly values, per EN 50160:1999[10] and CIGRE C4.07/CIRED.[7] Average values of this parameter are then compared with percent of Uh from IEC 61000-3-6: emission limits.[5]

5.3 MEASUREMENT PROCEDURE

5.3.1 Equipment

The process demands that recording instruments as well as voltage and current transducers comply with certain characteristics to assure that representative samples will be obtained.[3] The analog input bandwidth relates to the frequency limit above which the signal is attenuated by more than 3 dB (29.2%). IEEE-519[3] recommends that the bandwidth of 3 ± 0.5 Hz between the −3dB points with a minimum attenuation of 40 dB at a frequency of f_h + 15 Hz should be used. A 1.5-kHz analog input bandwidth would limit the harmonic measurement up to the 25th harmonic in a 60-Hz and to the 30th in a 50-Hz system. This covers most frequencies of interest in practical applications. Considering Nyquist criterion, if the input signal contains frequencies higher than half the sampling frequency, the signal cannot be correctly interpreted and an analog input bandwidth greater than 3 kHz will be required.

For all harmonic currents below the 65th (3.9 kHz in a 60-Hz or 3.25 kHz in a 50-Hz system) to be processed properly, the sampling frequency should be at least twice the desired input bandwidth, or 8 k samples per second in this case, to cover 50- and 60-Hz systems. The requirement is for 95% or better accuracy and minimum required attenuation of 50 to 60 dB for 30-Hz; 30 to 50 dB for 120- to 720-Hz; 20 to 40 dB for 720- to 1200-Hz; and 15 to 35 dB for 1200- to 2400-Hz signals. The lower limit is for frequency domain and the higher limit is for time domain instruments.[3] These limits have to do with the attenuation of high-frequency signals when the instrument is tuned at the fundamental frequency.

A large variety of instrumentation exists that can be used to carry out measurements and long-term recordings. Power quality analyzers are capable of carrying out measurements of rms voltage and current and perform calculations of active, reactive, and apparent power. They also compute harmonic distortion of voltage and current signals presenting individual and total harmonic levels, and some of them can calculate V*t and I*t products and K factor. There are indoor and outdoor versions of monitoring equipment and some of them can be set up to carry out long-term recordings.

5.3.2 Transducers

These elements convert the parameter to measure in a signal of adequate amplitude to be processed by the measuring equipment. However, not only amplitude is important. It is essential that their frequency response have an appropriate bandwidth so as not to produce any signal distortion. As transducers that comply with these requirements, the following can be utilized:

Potential transformers (PT)
Current transformers (CT)

Depending on the system voltage and the network configuration and type of load, the voltage can be measured directly or through the PTs. With regard to current measurements, they can be carried out on the primary side using the current probes furnished with the measuring equipment or at the low-voltage side, usually at the utility meter location. Under uncertainty regarding their frequency response, transducers should be subjected to tests to determine that their bandwidth is adequate to carry out harmonic measurements.

Although IEEE-519[3] points out that most utility measuring TPs can be used with a precision of 97% in the frequency range up to around 5 kHz, it is recommended that tests be conducted on TPs to determine that their bandwidth is appropriate up to the frequency of interest. In the case of TCs (those installed at the substation by the power utility for electric current and watt-hour measurements), they have a frequency bandwidth up to 20 kHz with an error smaller that 3%, according to reference 3. Properly grounded (complying with IEEE 518-1992[11]) shielded coaxial cables are recommended for short distances to the measurement equipment. If distances are large over a few tens of meters, fiber optic links are highly recommended to avoid all types of interference on the sometimes small amplitudes' signals.

5.4 RELEVANT ASPECTS

Harmonics and flicker increase with the use of power electronic devices in the system and create problems for loads susceptible to power quality problems. Standards for power quality and measurements to locate the source of problems are needed to maintain power quality.

The observed practice in relation to the monitoring of electrical parameters shows that measurements should be carried out in at least the PCC and at nodes where nonlinear loads are connected. The former location is important because compliance with standards must be sought there and the latter to verify emission limits at nonlinear load locations. This can be done actually at network nodes that group a number of similar nonlinear loads or at individual points with a single, large-size harmonic-producing load. We should keep in mind that if harmonic filtering is considered an option in a large industrial facility, cost will often determine the location of the filters and thus the point to monitor. This allows verification of the efficacy of the filtering scheme applied.

If interharmonics (noninteger multiples of fundamental frequency found in cycloconverters, arc welders, and electric furnace applications) are of interest, power quality monitoring equipment with the adequate bandwidth and accuracy must be used.

Harmonics measurements at specific customer sites may provide valuable information to determine compliance of end users with standards. It should be borne in mind, though, that the proliferation of electronic switching practically makes every customer a contributor to the harmonic distortion problem. This only makes it somewhat more complicated for utilities to determine specific responsibilities in the case of noncompliance to limits. Exhaustive measurement campaigns at suspect customer sites involving simultaneous observation of parameters at several sites may be required. Determining the right time for and duration of such endeavors must be conducted following the recommended guidelines.

Multiple sources of harmonic distortion thus require that steps toward characterizing emission levels at specific areas of the network be undertaken by utilities and industry as an effort to obtain an overall picture of potential trouble areas. This need will become increasingly evident as the networks expand and conducting surveys at specific locations turns progressively burdensome. Statistical prediction methods and analysis may need to be used as a helpful tool in this process.

If subharmonics are of interest, proper equipment should be used. Interharmonic limits should be limited below harmonic components. Measuring subharmonics is a challenging task.[12] The reason for this is that the frequency range up to 100 Hz remains very sensitive to spectral leakage problems caused by small synchronization errors; measuring subharmonics should be undertaken with this in mind.

REFERENCES

1. Arseneau, R., The performance of demand meters under varying load conditions, *IEEE Trans. Power Delivery*, vol. (4), October 1993.

2. Arseneau, R. and Filipski, P., Application of a three-phase nonsinusoidal calibration system for testing energy and demand meters under simulated field conditions, *IEEE Trans. Power Delivery*, PWRD-3(3), 874–879, 1998.
3. ANSI/IEEE 519:1992, IEEE Recommended Practices and Requirements for Harmonic Control in Electrical Power Systems, IEEE, 1992.
4. IEC 61000-4-7 Electromagnetic compatibility (EMC) — part 4-7: Testing and measurement techniques — general guide on harmonics and interharmonics measurements and instrumentation, for power supply systems and equipment connected thereto, ed. 2, 2002.
5. IEC 61000-3-6, Assessment of emission limits for distorting loads in MV and HV power systems, technical report type 3, 1996.
6. van den Bergh, M., Harmonics and flicker requirements and instrumentation, *Conformity*, August 2004, Input #22.
7. Joint WG CIGRE C4.07/CIRED, Power quality indices and objectives, final WG report, January 2004, rev. March 2004.
8. IEC 61000-3-2 Electromagnetic compatibility (EMC) — part 3-2: limits — limits for harmonic current emissions (equipment input current ≤ 16 A per phase), 2001-10.
9. IEC 61000-2-2, Electromagnetic compatibility — part 2-2, environment compatibility levels for low-frequency conducted disturbances and signaling in public and low-voltage power supply systems, 2002.
10. CENELEC EN 50160:-1999, Voltage characteristics of electricity supplied by public distribution systems. European standard (supersedes 1994 ed.).
11. IEEE Standard 518-1982, IEEE Guide for the Installation of Electrical Equipment to Minimize Electrical Noise Inputs to Controllers from External Sources.
12. Testa, A. and Langella, R., Power system subharmonics, *Proc. 2005 IEEE Power Eng. Soc. Gen. Meet.*, June 12–16, 2005, San Francisco, CA.

6 Harmonic Filtering Techniques

6.1 INTRODUCTION

In a general context, we can refer to harmonic filters as passive and active filters. Their essential difference, as illustrated later, stands on whether they provide a (passive) filtering action within a selected bandwidth or as a result of a real-time (active) monitoring process that leads to the injection of real-time canceling harmonic currents.

One of the most common methods for control of harmonic distortion in industry is the use of passive filtering techniques that make use of single-tuned or band-pass filters. Passive harmonic filters can be designed as single-tuned elements that provide a low impedance path to harmonic currents at a punctual frequency or as band-pass devices that can filter harmonics over a certain frequency bandwidth.

The more sophisticated active filtering concepts operate in a wide frequency range, adjusting their operation to the resultant harmonic spectrum. In this way, they are designed to inject harmonic currents to counterbalance existing harmonic components as they show up in the distribution system. Active filters comprise CD, CA, series, and parallel configurations. Hybrid filters are a combination of passive and active filtering schemes. Active filtering is so extensive and specialized that it is not possible to cover it within the scope of this book.

This chapter presents a straightforward methodology to design a passive filter based on the relationship between fundamental parameters. It also makes use of IEEE guidelines[1,2] for the selection of the filter components, presenting some application examples, some of which use commercial software for harmonic filter analysis.

6.2 GENERAL ASPECTS IN THE DESIGN OF PASSIVE HARMONIC FILTERS

Passive filters are the most commonly used filters in industry. As illustrated in Figure 6.1, the following can be found under this category:

Single-tuned filters
High- (or band-) pass filters (first, second, and third order)

FIGURE 6.1 Electric diagrams of passive filters.

6.3 SINGLE-TUNED FILTERS

Probably the most common harmonic filter in industrial applications, the passive filter presents very low impedance at the tuning frequency, through which all current of that particular frequency will be diverted. Thus, passive filter design must take into account expected growth in harmonic current sources or load reconfiguration because it can otherwise be exposed to overloading, which can rapidly develop into extreme overheating and thermal breakdown. The design of a passive filter requires a precise knowledge of the harmonic-producing load and of the power system. A great deal of simulation work is often required to test its performance under varying load conditions or changes in the topology of the network.

Because passive filters always provide reactive compensation to a degree dictated by the voltampere size and voltage of the capacitor bank used, they can in fact be designed for the double purpose of providing the filtering action and compensating power factor to the desired level. If more than one filter is used — for example, sets of 5th and 7th or 11th and 13th branches — it will be important to remember that all of them will provide a certain amount of reactive compensation.

As discussed earlier, this filter is a series combination of an inductance and a capacitance. In reality, in the absence of a physically designed resistor, there will always be a series resistance, which is the intrinsic resistance of the series reactor sometimes used as a means to avoid filter overheating. All harmonic currents whose frequency coincides with that of the tuned filter will find a low impedance path through the filter.

The resonant frequency of this filter can be expressed by the following expression:

$$f_0 = \frac{1}{2\pi\sqrt{LC}} \tag{6.1}$$

where:
f_0 = resonant frequency in hertz
L = filter inductance in henrys
C = filter capacitance in farads

As later described in Section 6.3.3, the quality factor, Q_f, of the filter is the ratio between the inductive or capacitive reactance *under resonance* and the resistance.

Harmonic Filtering Techniques

Typical values of Q_f fluctuate between 15 and 80 for filters that are used in industry. The following relation is used to calculate the quality factor:

$$Q_f = \frac{X_L}{R} = \frac{X_C}{R} \qquad (6.2)$$

Low-voltage filters (480 to 600 V) use iron cores with air gaps that have elevated losses but are associated to low Q_f values. Medium–voltage filters (4.16 to 13.8 kV) have Q_f values in the upper range.

The relation between X/R of low-voltage systems ranges between 3 and 7. These systems do not present an elevated parallel resonant peak in the Z–f characteristic. Although low-voltage filters have elevated losses, they also provide greater attenuation to any oscillation present in the system.

The process of designing a filter is a compromise among several factors: low maintenance, economy, and reliability. The design of the simplest filter that does the desired job is what will be sought in the majority of cases.

The steps to set up a harmonic filter using basic relationships to allow for a reliable operation can be summarized as follows:

1. Calculate the value of the capacitance needed to improve the power factor and to eliminate any penalty by the electric power company. Power factor compensation is generally applied to raise power factor to around 0.95 or higher.
2. Choose a reactor to tune the series capacitor to the desired harmonic frequency. For example, in a six-pulse converter, this would start at the fifth harmonic and it would involve lower frequencies in an arc furnace application, as illustrated by the harmonic spectrum of Figure 2.20.
3. Calculate the peak voltage at the capacitor terminals and the rms reactor current.
4. Choose standard components for the filter and verify filter performance to assure that capacitor components will operate within IEEE-18[2] recommended limits. This may require a number of iterations until desired reduction of harmonic levels is achieved.

Passive filters carry a current that can be expressed as a fraction of the load current at fundamental frequency. As for their cost, they are more expensive than series reactors often used to provide some harmonic attenuation, but they have the advantage of providing reactive power at fundamental frequency. For practical purposes, they are substantially used in industry.

Filter designs usually offer a robust mechanism that provides some minor filtering action for a fraction of other harmonic currents whose order is close to the tuning frequency, provided that no filters tuned at those frequencies exist.

Filter impedance must be smaller than that presented by the system at the tuning frequency. In low-voltage systems in which ratio X/R is small, an individual filter may be sufficient to provide the necessary attenuation. For example, neglecting the

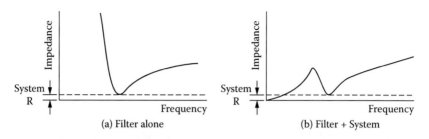

FIGURE 6.2 Frequency response of a single-tuned harmonic filter.

intrinsic resistance of the series reactor in a harmonic filter, the lowest value of the impedance frequency characteristic in Figure 6.2, as seen from the source, results from the resistive component of the system. The location of this point on the y-axis at the tuning frequency would be around three times higher for a network with an X/R ratio of 10 as compared with a case in which X/R is equal to 3. A resistive component with a theoretical zero resistance would make the filter absorb the entire harmonic current of frequency equal to the tuning frequency of the filter. Sometimes a series resistive component is included to control the maximum current allowed through the filter. This will have an impact on the quality factor of the filter as described by Equation (6.2).

The study of the response of single-tuned filters reveals the following relevant aspects:

> Single-tuned filters act as a small impedance path, effectively absorbing the harmonic currents for which they are tuned. It is important to be aware that these filters may take currents of neighboring frequencies, as will be shown in this chapter. Therefore, it will be important to assess the impact of those additional currents on the operational parameters of the filter.
>
> For a typical power utility or industrial system, there is a pronounced increase in the impedance slightly below the series LC filter tuning frequency as shown in Figure 6.2(b). This is a result of a parallel resonant condition between the capacitor of the filter and the inductance of the source.

6.3.1 Design Equations for the Single-Tuned Filter

The impedance of the filter branch is given by:

$$Z = R + j\left[\omega L - \frac{1}{\omega C}\right] \quad (6.3)$$

where R, L, and C are the resistance, inductance, and capacitance of the filter elements, respectively, and ω is the angular frequency of the power system.

The series resonance condition is excited when the imaginary part of the impedance is equal to zero, where the only impedance component left is the resistance. The frequency at which the filter is tuned is then defined by the value of ω that

Harmonic Filtering Techniques

makes inductive and capacitive reactance cancel one another in Equation (6.3). This frequency is given by Equation (6.1). If we make h the ratio between the harmonic and the fundamental frequencies of the system, the inductive and capacitive reactances at the harmonic frequency can be expressed as:

$$X_{Lh} = h\omega_L \tag{6.4}$$

$$X_{Ch} = \frac{1}{(h\omega_C)} \tag{6.5}$$

Expressed in a different way, assuming zero resistance, the condition for the impedance in Equation (6.3) dropping to zero at the tuning frequency requires:

$$X_{Lh} = X_{Ch} \tag{6.6}$$

Substituting Equation (6.4) and Equation (6.5) in Equation (6.6) and solving for h, we get:

$$h^2 = \frac{X_C}{X_L}$$

or

$$h = \sqrt{\frac{X_C}{X_L}} \tag{6.7}$$

6.3.2 Parallel Resonant Points

As mentioned earlier, the interaction of the filter with the source impedance (Ls) always results in a parallel resonance characterized by the large impedance peak illustrated in Figure 6.3. Seen from the capacitor bank upstream (toward the source), a parallel resonance is to be established at a frequency:

$$f\_par\_res = \frac{1}{2\pi\sqrt{(Ls+L)C}} \tag{6.8}$$

which falls slightly below the filter-tuning frequency, as illustrated in Figure 6.3.

In installations in which multiple single-tuned filters are required, a parallel resonant frequency will exist for every individual passive filter. Notice that the frequency of the parallel resonant point would experience a shift whenever changes in filter elements L or C or in source inductance Ls occurs. Ls can change, for

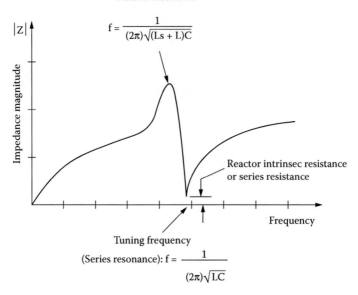

FIGURE 6.3 Resonant points on a single-tuned series RLC filter.

instance, following the disconnection or addition of a transformer at the substation. This could take place every time the power utility changed the configuration of source during transformer maintenance actions or whenever transformers are added to the bank.

A change that can also affect the parallel resonant frequency in Equation (6.8) is the addition of power factor capacitor units on the feeder that serves the nonlinear load where the harmonic filter is installed.

However, the most notable impact of the source impedance on the filter performance is its parallel impedance peak. Figure 6.4 illustrates the impedance value at parallel resonant peaks in a two-branch filter design as a function of source impedance. This is shown in three different plots with increasing short-circuit megavoltampere values (or reduced source impedance) in Figure 6.4. Observe how the parallel impedance peaks are reduced in amplitude, which will have the effect of decreasing harmonic distortion produced by any harmonic component of frequency smaller than

Harmonic Filtering Techniques

FIGURE 6.4 Response of a harmonic filter for different source MVAsc.

the tuning frequency of the filter. The opposite effect would be observed if we start with a small source impedance and increase it when a substation transformer is out for maintenance.

A problem that can arise with the adjacent parallel-resonant points is a detuning action. If the filter is tuned at exactly the frequency of interest, then a shifting of the series-resonant point to higher frequency values will result in a sharp impedance increase as seen by the harmonic current of that frequency order. This can occur, for example, from capacitor aging, which would cause some decrease in capacitance. If the parallel-resonance peak shifts in such a way that it aligns with the frequency of a characteristic harmonic of the load, the resultant harmonic voltage amplification can be disastrous because it can produce overvoltage stresses on solid insulation of cables and on machine windings. The aspects involved in the detuning action of the filter can be described as follows:

The tripping action of capacitor bank fuses disconnecting one or two single-phase units will decrease the equivalent three-phase capacitance, increasing the tuning frequency of the filter.

Manufacturing tolerances of the filter elements can result in a shifting of the tuning frequency in any direction, for which it is important to take them into account.

Temperature variations can produce an accelerated aging on the capacitor units.

The variations on the topology or configuration of the system, which changes the upstream inductive reactance seen from the location of the filter, can also have an impact on the location of parallel resonant points.

Considering the preceding points and assuming a fixed source impedance it is convenient to tune the filter at a frequency slightly below the desired frequency, typically 3 to 5%. This will account for small tune frequency shifts to higher values over time, yet allow the filter to provide a low impedance path. Also, an unbalance detection scheme to protect the capacitor bank and to assure the proper operation of the filter will be important to consider.

6.3.3 Quality Factor

Regarding single-tuned harmonic filters, the quality factor relates the ability of a filter to dissipate the absorbed energy at the tuned frequency. IEEE[1] quotes an approximate equivalent definition as the ratio of the resonant frequency, ω_θ, to the bandwidth between those frequencies on opposite sides of the resonant frequency where the response of the resonant structure differs 3 dB from that at resonance:

$$Q = \frac{\omega_\theta}{(\omega_1 - \omega_2)} \tag{6.9}$$

In an RLC series circuit, we can define Q as:

$$Q = \frac{1}{R}\sqrt{\frac{L}{C}} = \frac{X_{Lh}}{R} = \frac{X_{Ch}}{R} \tag{6.10}$$

where X_{Lh} and X_{ch} are the inductive and capacitive reactance, respectively, at the resonant frequency of the series filter. Figure 6.5 and Table 6.1 summarize the basic parameters that describe the single-tuned passive filter.[3]

FIGURE 6.5 Fundamental quantities that determine the performance of a passive single-tuned filter.

TABLE 6.1
Relevant Quantities on a Passive Single-Tuned Filter

Tuned Harmonic Order	Quality Factor	Bandwidth	Reactive Power at f_1	Active Power at f_1 (losses)
$h = \dfrac{f_n}{f_1} = \sqrt{\dfrac{X_C}{X_L}}$	$Q_f = \dfrac{n \cdot X_L}{R} = \dfrac{X_C}{n \cdot R}$	$B = \dfrac{f_n}{Q_f}$	$Q_C = \dfrac{V^2}{X_C} \cdot \dfrac{n^2}{(n^2 - 1)}$	$P \cong Q_C \cdot \dfrac{n}{n^2 - 1} \cdot \dfrac{1}{Q_f}$

Note: f_1 = fundamental frequency; $\omega = 2\pi f_1$ = angular frequency; f_n = tuning frequency; n = harmonic order f_n/f_1; V = nominal line-to-line voltage; X_L = inductor reactance at fundamental frequency = $L\omega$; and X_C = capacitor reactance at fundamental frequency = $1/\omega C$.

Figure 6.6 shows a number of plots for a harmonic filter with different Q_f values. The shaded area delineates the response of the filter. Notice how the larger the Q_f, the better the filtering action achieved, which is reflected on the lowest impedance at the tuning frequency. This is a logical effect that results from decreasing the purely resistive impedance at the resonant frequency in Equation (6.10).

The following points summarize the most relevant quality factor aspects in single-tuned filters:

> Typically, the resistance of a single-tuned harmonic filter is the intrinsic resistance of the reactor.
> However, R can be favorably used to vary the quality factor of the filter and provide a way to control the amount of desired harmonic current through it.
> A large Q_f value implies a prominent valley at the resonant (tuning) frequency of a filter and therefore the trapping of the largest amount of harmonic frequency.
> The best reduction of harmonic distortion will be achieved with large Q_f value filters. However, care should be exercised in assessing harmonic currents of frequencies other than the one for which the filter is tuned because they will also find a reduced impedance path. These currents will provide increased heat dissipation. It will often be necessary to conduct computer-aided harmonic simulation studies to predict the performance of the filters, especially when multiple harmonic sources exist.
> Lower quality factor filters could be used in situations in which harmonic distortion barely exceeds the limits and a small filtering action is all that is needed to bring it into compliance.

6.3.4 Recommended Operation Values for Filter Components

6.3.4.1 Capacitors

Capacitor banks are voltage-sensitive components of filters for the following reasons:

Capacitors may be exposed to increased voltage during waveform distortion produced by harmonic components.

The voltage increase can be in the form of an augmented peak or an increase of the rms value.

The reactive power output of the capacitor will change with the square of the rms-distorted to the rms-undistorted voltage ratio.

In situations in which switching overvoltages are frequent, filter capacitors can be exposed to damage

IEEE-18[2] recommends limits for the operation of shunt capacitors in power systems including current, power, and voltage across the capacitor units. Following this guideline, capacitors are designed to be operated at or below their rated voltage and to be capable of continuous operation under contingency system and bank conditions, provided that none of the limitations of Table 6.2 are exceeded.

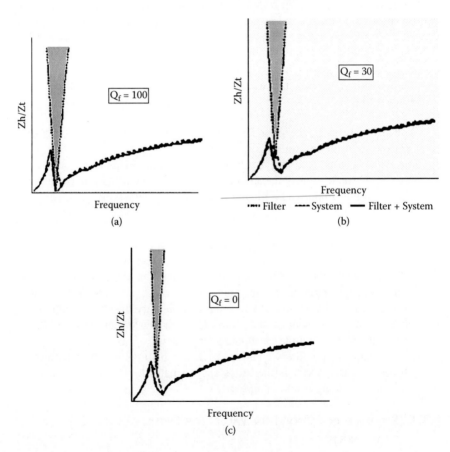

FIGURE 6.6 Impedance-frequency plots for a harmonic filter for different values of Q.

Harmonic Filtering Techniques

TABLE 6.2
Maximum Recommended Limits for Continuous Operation of Shunt Capacitors under Contingency Conditions

VAR	135%
RMS voltage	110%
Rated voltage, including harmonics	120%
RMS current	135%

Note that the limit for the rms current is lower that the 180% that was considered in the 1992 revision of IEEE-18 because that current level may be causing a rated power exceeding the recommended limit. It is important to observe compliance with these limits mainly at facilities at which considerable harmonic waveform distortion exists or capacitors are part of a harmonic filter.

Generally, the capacitor voltage in a bank used in a single-tuned harmonic filter will be exposed to an increased voltage that can be approximated in terms of the harmonic order (h) of the filter as follows:

$$V_{cap} = \frac{h^2}{(h^2-1)}(V_{system}) \tag{6.11}$$

When the maximum amplification of the voltage is verified, the worst conditions that include the maximum voltage considering the tolerance of the filter elements (typically 8% for the capacitor and 5% for the inductor) must be tested.

When a capacitor bank of a nominal voltage different from that of the system is used, the effective bank, $kVAR$, must be determined from the following expression:

$$kVAR_{effective} = \left(\frac{V_{system L-L}}{V_{cap L-L}}\right)^2 (kVA_{rated}) \tag{6.12}$$

The presence of a reactor in the filter changes the effective $kVAR$ of the filter. The new output will be:

$$kVAR_{filter} = \frac{V_{capL-L}}{(Z_C - Z_L)} \tag{6.13}$$

If reactive compensation is needed, the capacitor bank of the filter can be chosen to provide it. However, the designer will typically need to follow an iterative process to decide the suitable VARs of the bank.

6.3.4.2 Tuning Reactor

The maximum voltage elevation across the reactor must also be determined. The parameters usually included in the specification of the reactor are the following:

50-/60-Hz current
Harmonic current spectrum
Short-circuit current
X/R ratio
System voltage
BIL (basic insulation level)

Reactors used in harmonic filters are sometimes designed with an air core. This provides linear characteristics regarding frequency and current. In applications that involve industrial power systems, 5% in tolerance is typically used. The relation, X/R, at 60 Hz, which is typically smaller that 150, can be further manipulated to obtain the desired quality factor as described previously.

Also, the maximum voltage elevation across the reactor must be determined. The nominal voltage of the reactor must be able to handle the overvoltage imposed under a short-circuit condition, for instance, when a capacitor fails. The basic insulation level (BIL) of the reactor and, similarly, of the capacitor bank must be the same as that of the power transformer that feeds the load where the filters are integrated.

6.3.5 UNBALANCE DETECTION

The purpose of an unbalance detection scheme is to remove capacitor banks as soon as phase overcurrent protection trips due to a single fault-to-ground event. Generally, unbalance detection triggers an alarm when one or several capacitor stages are lost in the bank. In a harmonic filter, the failure of a capacitor unit can detune the filter and produce harmonic voltage amplification following a shift in the parallel resonant point.

6.3.6 FILTER SELECTION AND PERFORMANCE ASSESSMENT

> First, determine if reactive compensation is required. If this is the case, the capacitor bank of the filter must be sized to provide the needed VARs.
> For certain system conditions, more than one filter may be needed. Consider all possible scenarios to determine the worst-case condition. In certain applications, nonlinear loads may be cyclic and filter schemes must be designed so that they allow the possibility of having filter branch components in and out, as needed.
> When analyzing the effectiveness of a filter (understood as the degree of harmonic suppression), it is important to try it for different upstream impedance conditions, which are tied with the utility source impedance.
> Another important point to consider is the possibility of shifting the parallel resonant points. As described previously, changing feeder capacitance as a

Harmonic Filtering Techniques

response to load variations or voltage profile can have an impact on parallel resonant frequency. For instance, if at 100% feeder loading, there is a system resonant peak at the 4.5th harmonic, a loading decrease accompanied by disconnection of power factor capacitor banks may shift the parallel resonant peak to the 5th harmonic. Any fifth harmonic current from the load would see a large upstream impedance and a higher voltage harmonic distortion at that frequency would develop. This situation must be considered in the design of harmonic filters or in the implementation of special operation rules to minimize the negative effects of the resonant peak.

- The traditional design criteria in relation with the presence of harmonic currents in industrial networks are total harmonic distortion (THD) and telephone interference factor (TIF) levels. Harmonic distortion is likely to be exceeded in industrial and commercial applications involving large power converters, massive amounts of fluorescent lighting, and significant amounts of office equipment. Therefore, commercial and financial facilities, office or commercial buildings, and corporate and public offices are good candidates to exceed compliance with recommended limits. Likewise, telephone interference factor may reach considerable levels in extensive power cable networks under resonance conditions.
- THD should be evaluated at every relevant bus in the system — namely, at the main plant substation bus, at those nodes with harmonic current sources, and wherever sensitive equipment exists. If THD limits are above limits, then the need to provide harmonic filters must be sought and THD/TIF levels reassessed.
- Operation scenarios to be considered in the filter design stage should include network and load reconfiguration that involve plant expansion and future load growth.
- If the rated values of the filter components are exceeded under normal operation, an adjusting action should follow. However, any adjustment performed to the filter scheme should be anticipated and considered in the filter design specifications.
- The evaluation of harmonic filters must include power frequency and harmonic losses. This is particularly relevant for the design of a "minimum filter," i.e., that which is specified and installed to bring harmonic current distortion within limits but not for power correction purposes.

6.4 BAND-PASS FILTERS

Band-pass filters, high-pass in particular, are known by their small impedance value above the corner frequency. Typical frequency response of a high-band pass filter is shown in Figure 6.7. This filter draws a considerable percentage of frequency harmonic currents above the corner frequency. Therefore, this frequency must be placed below all harmonic currents that have an important presence in the installation.

In planning to adopt a high-pass filter as a harmonic mitigating measure, the following aspects should be considered:

FIGURE 6.7 Response of a high-pass filter for different Q_f values.

The impedance-frequency characteristic of a high-pass filter will entail a very different filtering action as compared with that provided by a single-tuned filter. See Figure 6.1 for the discussion that follows.

Harmonic current elimination using a high-pass filter may require a quite different sizing of filter elements, particularly of the capacitor bank, compared with a single-tuned filter. For example, a 3-MVAR bank used in a fifth harmonic filter in a 60-Hz application may fall short in size when used as part of a high-pass filter with a corner frequency of 300 Hz. Obviously, this will very much depend on the additional harmonic currents that the high-pass filter will be draining off. First-order high-pass filters are characterized by large power losses at fundamental frequency, for which they are less common.

The second-order high-pass filter is the simplest to apply; it provides a fairly good filtering action and reduces energy losses at fundamental frequency.

The third-order high-pass filter presents greater operating losses than the second-order high-pass filter and is less effective in its filtering action.

The impedance of the second-order high-pass filter can be expressed as:

$$Z = \frac{1}{j\omega C} + \frac{1}{\left(\frac{1}{R} + \frac{1}{j\omega L}\right)} \qquad (6.14)$$

The corner frequency of the filter is given by:

Harmonic Filtering Techniques

$$f = \frac{1}{2\pi\sqrt{LC}} \qquad (6.15)$$

The quality factor of the filter is calculated as:

$$Q_f = \frac{R}{\sqrt{\frac{L}{C}}} = \frac{R}{X_L} = \frac{R}{X_C} \qquad (6.16)$$

For typical high-pass filters, Q_f values between 0.5 and 2.0 are found. Filters with higher Q_f would provide a superior filtering action at the corner frequency, although at higher frequencies, the impedance would show a steady increase, as illustrated in Figure 6.7. Filters with smaller Q_f values would show an inferior performance at the corner frequency, although at frequencies higher than that, a less pronounced increase of impedance with frequency is obtained. This is also illustrated in Figure 6.7.

Other factors that must be considered in the selection of Q_f are the following:

The tuning frequency of the filter
Concerns for telephone interference (if it exists)
Power losses

6.5 RELEVANT ASPECTS TO CONSIDER IN THE DESIGN OF PASSIVE FILTERS

A number of aspects must be considered in the design stage of passive filters for controlling problems associated with harmonics. These are summarized as follows:

The capacitive kVAR requirements for power factor correction. Some installations may benefit from the installation of harmonic filters because power factor will be improved. In other situations, power factor correction needs may dictate the size of the capacitor bank to use.
In single-tuned filters, watch the resonant parallel peaks resulting from the interaction between the filter and the source.
Consider tolerances of filter components. They may produce undesirable shifts of resonance frequencies.
Also look for load and network impedance changes that may modify established worst-case harmonic scenarios.
Oversized capacitor banks may be required in high-pass filters with low corner frequencies and significant higher order characteristic harmonics.
Be aware of quality factor filters as a measure to control the amount of harmonic currents to be drawn from the system. Avoid overloading capacitor banks using a series resistor in single-tuned filters. A trade-off between

decreased THD values and power factor correction assuring capacitor bank integrity will often decide the Q_f value to adopt in a filter.

Extensive electric networks may have nonlinear loads with different spectral content. Whenever possible, grouping loads by type of harmonic spectrum (for instance, 6-pulse converters, 12-pulse converters, arcing type devices, fluorescent lighting, etc.) can optimize the installation, location, and sizing of harmonic filters. Although this is a difficult task to achieve, especially when comparable types of loads are not on the same location, the idea should be considered as a way to reduce the number of harmonic filters to install. Load grouping could also help reduce telephone interference by trying to keep telephone lines as distant as possible from sites carrying higher-order harmonic currents.

Minimum filters may be adopted under no reactive compensation needs. The parameters of a minimum filter must be chosen to reach the maximum recommended THD limit.

Always watch for filter power losses.

6.6 METHODOLOGY FOR DESIGN OF TUNED HARMONIC FILTERS

The recommended procedure for the design and validation of single-tuned harmonic filters is summarized in the following sections.

6.6.1 SELECT CAPACITOR BANK NEEDED TO IMPROVE THE POWER FACTOR FROM THE PRESENT LEVEL TYPICALLY TO AROUND 0.9 TO 0.95

The capacitive reactance needed to compensate the needed VARs to improve the power factor from PF_1 (associated with θ_1) to PF_2 (associated with θ_2) is given by:

$$VARs = P (\tan \theta_2 - \tan \theta_1) \tag{6.17}$$

with

$$P = (V)(I) \cos \theta_2 \tag{6.18}$$

Sometimes P must be calculated from multiplying the apparent power, S, by the power factor of the load.

The capacitive reactance required is obtained with the following relation:

$$X_{C1} = \frac{V^2}{VARs} \tag{6.19}$$

where V and $VARs$ are capacitor-rated values.

Harmonic Filtering Techniques

At harmonic frequency h, this reactance is:

$$X_{Ch} = \left(\frac{1}{h}\right) X_{C1} \quad (6.20)$$

6.6.2 Choose Reactor that, in Series with Capacitor, Tunes Filter to Desired Harmonic Frequency

The inductive reactance required at harmonic h is, in this case:

$$X_{Lh} = X_{ch} \quad (6.21)$$

and at fundamental frequency, it is:

$$X_{L1} = \left(\frac{1}{h}\right) X_{Lh} \quad (6.22)$$

6.6.3 Determine Whether Capacitor-Operating Parameters Fall within IEEE-18[2] Maximum Recommended Limits

6.6.3.1 Capacitor Voltage

The rms and peak voltage of the capacitor must not exceed 110 and 120%, respectively, of the rated voltage. They can be determined as follows:

$$V_{C_{rms}} = \sqrt{(V_{C1}^2 + V_{Ch}^2)} \quad (6.23)$$

$$V_{C_{peak}} = \sqrt{2}(V_{C1} + V_{Ch}) \quad (6.24)$$

where voltage through the capacitor at fundamental frequency is given by:

$$V_{C1} = X_{C1} I_{C1} \quad (6.25)$$

I_{C1} is the current through the capacitor and it is calculated in terms of the maximum phase-to-neutral voltage, which in turn is specified 5% above the rated value, to account for voltage regulation practices:

$$I_{C1} = (1.05) \frac{V_{L-N}}{(X_{C1} - X_{L1})} = (1.05) \left[\frac{\left(\frac{V_{L-L}}{\sqrt{3}}\right)}{(X_{C1} - X_{L1})} \right] \quad (6.26)$$

V_{ch} is found in terms of I_{ch}, which must be determined from measurements or from a typical harmonic spectrum of the corresponding nonlinear load.

$$V_{ch} = X_{ch} I_{ch} \tag{6.27}$$

6.6.3.2 Current through the Capacitor Bank

The RMS current through the capacitor bank must be within 135% of the rated capacitor current, to comply with IEEE-18. Its value is determined from the fundamental current and from the harmonic currents under consideration:

$$I_{Crms} = \sqrt{(I_{C1}^2 + I_{Ch}^2)} \tag{6.28}$$

6.6.3.3 Determine the Capacitor Bank Duty and Verify that It Is within Recommended IEEE-18 Limits

$$kVAR = \frac{(V_{Crms})(I_{Crms})}{1000} \tag{6.29}$$

where V_{Crms} is the voltage through the capacitor calculated in Equation (6.23) and I_{Crms} is the current through the capacitor of Equation (6.28).

If IEEE-18 is not met, the process may require more than one iteration resizing the size of the capacitor bank.

6.6.4 TEST OUT RESONANT CONDITIONS

Once the filter parameters have been selected, it is important to verify that nonresonant conditions are presented between the capacitor bank of the filter and the inductive reactance of the system. To carry out this task rigorously, a harmonic analysis program is needed to determine the frequency response of the system and to assess whether the desired reduction in harmonic distortion levels is achieved. We show next how far we can go by manually applying the described procedure.

6.7 EXAMPLE 1: ADAPTATION OF A POWER FACTOR CAPACITOR BANK INTO A FIFTH HARMONIC FILTER

Suppose that a capacitor bank installed for reactive power compensation at a six-pulse power converter application is to be tuned to the fifth harmonic. We need to determine the required reactor size and verify whether capacitor bank operation parameters fall within IEEE-18 recommended limits.

Harmonic Filtering Techniques

Assumed data:

Harmonic current to filter: fifth
System phase-to-phase voltage: 13.8 kV
Power factor capacitor bank size: 4.5 MVAR @ 15 kV.
Plant load: 8 MVA composed of six-pulse static power converters

Using the preceding methodology, we carry out the following calculations.

Capacitor bank reactance. Typically, the X/R relation for this type of bank is of the order of 5000; therefore, the resistance can be ignored.

$$X_{C1} = \frac{kV_{L-L\,rated}^{2}}{MVAR_{rated}} = \frac{15^2}{4.5} = 50 \quad [Ohms]$$

Calculate the series reactor required. Air-core reactors typically have an X/R ratio of the order of 30–80. Again, resistance can be disregarded.

From Equation (6.20) through Equation (6.22) we obtain:

$$X_{L1} = \frac{X_C}{h^2} = \frac{50}{5^2} = 2.0 \quad [Ohms]$$

Determine whether capacitor-operating parameters fall within IEEE-18 recommended limits. rms current through the filter:

$$I_1 = \frac{V_{L-N}}{(X_C - X_L)} = \frac{(1.05)\left(\frac{13,800}{\sqrt{3}}\right)}{(50-2)} = 174.3 \quad [A]$$

If we assume the harmonic current from the load is inversely proportional to the fundamental current:

$$I_5 = \frac{1}{h}\frac{kVA_{load}}{\sqrt{3}(13.8)} = \frac{1}{5}\left(\frac{8000}{\sqrt{3}(13.8)}\right) = 66.9 \quad [A]$$

$$X_{C5} = \frac{X_{C1}}{h} = \frac{50}{5} = 10 \quad [\Omega]$$

Peak and rms voltage through the capacitor:

$$V_{C\,peak} = \sqrt{2}(V_{C1} + V_{Ch}) = \sqrt{2}(X_{C1}I_1 + X_{C5}I_5)$$

$$= \sqrt{2}(50 \times 174.3 + 10 \times 66.9)$$

$$= \sqrt{2}(8715 + 669) = 13{,}271\ [V]$$

$$V_{C\,rms} = \sqrt{(V_{C1}^2 + V_{Ch}^2)} = \sqrt{(8715^2 + 669^2)} = 8741\ [V]$$

If 8660-V (line-to-neutral voltage for a 15-kV system) capacitors are used, then the capacitor voltage is as follows:

$$\frac{V_{C\,rms}}{V_{C\,rated}} = \frac{8{,}741}{8{,}660} = 1.009\,p.u.$$

(Below the 1.1 p.u. limit of IEEE-18)

$$\frac{V_{C\,peak}}{V_{C\,peak\,rated}} = \frac{13{,}271}{\sqrt{2}(8{,}660)} = 1.084\,p.u.$$

(Below the 1.2 p.u. limit of IEEE-18)

The rms current through the reactor is the summation of all rms currents that will flow through the filter. The assumption here is that only the fifth harmonic is involved:

$$I_{C\,rms} = \sqrt{I_1^2 + I_5^2} = \sqrt{174.3^2 + 66.9^2} = 186.7\ [A]$$

$$\frac{I_{C\,rms}}{I_{C\,rated}} = \frac{186.7}{\left(\dfrac{4500}{\sqrt{3}(15)}\right)} = \frac{186.7}{173} = 1.08$$

(Below the 1.35 p.u. limit of IEEE-18)

However, caution should be exercised because a harmonic filter often serves as a sink for currents from adjacent frequencies. Without a harmonic load flow program, it is impossible to be precise about the amount of harmonic currents (other than those for which the filter is tuned) that will flow through it. Thus, in approximate calculations, a factor of 1.15 to 1.2 is sometimes used. In our example, we are within limits, even considering the largest range:

Harmonic Filtering Techniques

$$\text{Recalculated } \frac{I_{C\,rms}}{I_{C\,rated}} = \frac{186.7}{\left(\frac{4500}{\sqrt{3}(15)}\right)}(1.2) = \frac{186.7}{173} = 1.296 < 1.35$$

Reactive power delivered by the capacitor bank is:

$$kVAR \text{ per phase} = \frac{(V_{Crms})(I_{Crms})}{1000} = \frac{\sqrt{(V_{C1}^2 + V_{C5}^2)} \cdot \sqrt{(I_{C1}^2 + I_{C5})^2}}{1000}$$

$$= \frac{(8741)(186.7)}{1000} = 1632$$

Three-phase capacitor power is:

$$kVAR_{3\text{-phase}} = 1887 \times 3 = 4896 = 4.9 \text{ MVAR}$$

The total capacitor output will be derated because the capacitor bank is of a higher–voltage class:

$$MVAR_{derated} = MVA_{rated}\left(\frac{kV_{L-L}}{kV_{rated}}\right)^2 = 4.5\left(\frac{13.8}{15}\right)^2 = 3.8$$

Therefore, considering the influence of the load harmonics, the ratio between delivered and (de)rated power is:

$$\frac{kVAR_{3\text{-phase}}}{kVAR_{rated}} = \frac{4.9}{3.8} = 1.29$$

(Below the 1.35 p.u. limit of IEEE-18)

Up to this point, we would only need to verify the parallel resonant points. However, because the assumed load does not contain a characteristic harmonic of frequency lower than the fifth harmonic, it really does not matter where the parallel resonant lies.

6.8 EXAMPLE 2: DIGITAL SIMULATION OF SINGLE-TUNED HARMONIC FILTERS

This example illustrates that harmonic analysis is greatly simplified, on one hand, and that more valuable information is obtained for the assessment, on the other hand, using specialized harmonic analysis software. One of the pieces of information that

FIGURE 6.8 Electric diagram of an oil offshore installation with a harmonic-producing load.

is extremely helpful in the analysis of mitigating measures for harmonic control is the frequency-dependent plots. These include spectral content (harmonic spectrum) of voltage and current signals and impedance-frequency characteristics of the distribution system before and after the application of filters. Therefore, it is possible to assess the performance of the system step by step with the application of different mitigating methods like for instance increasing the number of harmonic filters until desired THD values are reached. This type of analysis is often combined with economical assessment to obtain a comprehensive evaluation harmonic filtering or any other harmonic control scheme.

Figure 6.8 shows a diagram of a typical installation of an electric submersible pump in an oil field offshore platform involving a variable frequency drive fed off from an individual synchronous generator. In these types of installations, in which all or most of the load is nonlinear, it is common to experience large waveform harmonic distortion due to the lack of linear load components, which act as natural attenuators of waveform distortion. Harmonic filters at the primary of the transformer that feeds the VFD are tested to show the reduction of harmonic waveform distortion at the generator terminals.

The filter capacitor banks are sized following an inverse relation to their harmonic order. For example, the 5th and 7th harmonic filters are chosen as one fifth and one seventh, respectively, of the converter load.

The results obtained for this example are presented in the form of current and impedance vs. frequency diagrams in Figure 6.9 through Figure 6.12.

Harmonic Filtering Techniques

FIGURE 6.9 Current in branch 1–2, with no filter, THDI = 28.1%.

Figure 6.9 shows the harmonic spectra of the current at branch 1 to 2 with no filters installed. Figure 6.10(a–c) describe the effect of a 100-kVA fifth harmonic filter including branch 1 to 2 and filter currents as well as the Z–f plot that portrays the series (Z = 0) and parallel (just prior to the series) resonant points of the filter. Figure 6.11(a–c) shows the same results for a scenario in which fifth and seventh filter branches are applied. Finally, Figure 6.12(a–d) shows the effect of additionally including the 11th harmonic branch.

Table 6.3 summarizes the THD under different filtering scenarios. Notice that the THD_V level without any filter at the VFD is already close to the IEEE-519 limit of 5%, while the THD_I is well above the 5% threshold for all power generation equipment. With the installation of the fifth harmonic filter, the THD_I is reduced to less than half its value when there are no filters, but it is still above the 5% IEEE threshold. If fifth and seventh harmonic filter branches are added, we reach the point at which THDV and THDI fall below the recommended limits of IEEE-519.

However, regarding IEEE-18 compliance,[2] the fifth harmonic filter is observed to result slightly above the recommended limits, as observed in Table 6.4. Notice in Figure 6.10(b), Figure 6.11(a), and Figure 6.12(a), how the current though the fifth harmonic filter, which initially comprises a fraction of other harmonics, becomes pure 5th harmonic when 7th- and 11th harmonic filters are added. However, when examined in terms of rms values, current remains practically unchanged. A similar behavior is observed for the resultant filter killivoltampere figures. The current and power values in Table 6.4 are those obtained considering the three filter

FIGURE 6.10 Fifth harmonic filter applied. *Continued.*

Harmonic Filtering Techniques

(c) Impedance-frequency curves

FIGURE 6.10 *Continued.*

branches connected. All capacitor peak voltage ratios resulted below the recommended 1.2 limit.

Therefore, a somewhat increased size for the fifth harmonic filter would bring all operating parameters under IEEE-18 compliance.

6.9 EXAMPLE 3: HIGH-PASS FILTER AT GENERATOR TERMINALS USED TO CONTROL A RESONANT CONDITION

This example is aimed at illustrating the reduction of THD_V using a high-pass filter in an installation similar to that of the oil company offshore platform of Example 2. The example describes another real-world application in which harmonic-related problems arose and even damaged a generator unit. This case presented an excellent opportunity for conducting investigations on a complicated phenomenon. The AC source, a 3-MVA synchronous turbo generator, was the power supply for two 1.5-MVA step-down transformers, which in turn fed a number of VFDs powering downhole electrosubmersible pumps. Apart from a small service transformer that sourced the platform services, the VFDs were the only loads as illustrated in Figure 6.13.

A number of harmonic measurements indicated that voltage and current harmonic levels were excessively high. The suspecting element was a 0.27-μF surge

FIGURE 6.11 Fifth- and seventh harmonic filters applied. *Continued.*

Harmonic Filtering Techniques

(c) Impedance-frequency curves with 5th and 7th harmonic filters

FIGURE 6.11 *Continued.*

(a) 5th harmonic filter current, Irms = 19A

FIGURE 6.12 Fifth, seventh, and eleventh harmonic filters applied. *Continued.*

(b) 7th harmonic filter current, Irms = 13A

(c) 11th harmonic filter current, Irms = 7A

FIGURE 6.12 *Continued.*

Harmonic Filtering Techniques

(d) Impedance-frequency curves with 5th 7th and 11th harmonic filters

FIGURE 6.12 *Continued.*

TABLE 6.3
Voltage and Current Harmonic Distortion at Generator Terminals

Order of Harmonic Filter Tested	No Filters	5th	5th and 7th	5th, 7th, and 11th
THDV (%)	4.54	1.9	0.98	0.46
THDI (%)	28.1	8.2	3.0	1.0

TABLE 6.4
Capacitor Bank Parameters Relative to IEEE-18

Harmonic Filter	Rated Current	RMS Current	$I_{RMS}\ I_{rated}$	Rated KVA	RMS kVA	$kVA_{RMS}\ kVA_{rated}$
5	13.9	19	1.37[a]	100	142	1.42[a]
7	9.7	13	1.34	70	95	1.21
11	6.2	7	1.13	45	55	0.91

[a] Above IEEE-18 recommended limits.

FIGURE 6.13 A high-pass filter at the generator bus on an offshore oil field installation.

protection capacitor bank, which apparently combined with connecting cables between generator and transformers to excite a resonant condition at the generator bus. Through additional measurements and simulation work, this was further confirmed to be the case. Here, we will show the resonant condition and how it was controlled using a high-pass filter to reduce harmonic distortion levels at the generator terminals within IEEE-519 limits.

Figure 6.14 shows the abrupt increase in impedance (dotted line) revealing a resonant condition around harmonics 39 through 43. Such a situation imposed severe stresses on the generator that comprised intense shaft vibration and increased operation temperature. Also shown is the Z–f characteristic of the band-pass filter that was applied at the 600-V generator bus. Filter elements were selected to obtain a corner frequency centered at around the 11th harmonic. Notice the system response with the high-pass filter showing a substantial reduction of the impedance at the resonant point.

Figure 6.15 shows how the impedance frequency characteristic looks at the primary of one of the downstream transformers feeding a VFD at one of the oil wells. At these locations, 5th-, 7th-, and 11th-harmonic filters were installed; this is noticeable on the Z–f characteristic, which also reflects the band-pass filter effect on the distribution system as seen from that location.

Thus, the installation of a low-cost high-pass filter (involving a 67-kVA capacitor bank) at the generator bus allowed a very annoying condition to be brought under control. Interestingly, the resonant phenomenon, as observed in Figure 6.14, involved a frequency range in which characteristic harmonics show very small (but apparently strong enough) values that, after undergoing amplification, made THD levels soar.

It is important to mention that applications of high-pass filters at the generator bus are usually combined with the application of single-tuned filters at the VFD locations to get THD levels within recommended limits at the generator terminals as well as at the individual VFD sites. Leaving the high-pass filter at the generator bus as the only

Harmonic Filtering Techniques 123

FIGURE 6.14 Impedance vs. frequency characteristics at the generator bus.

FIGURE 6.15 Impedance vs. frequency characteristics at a downstream VFD.

harmonic mitigating method may cause excessive heating on the high-pass filter elements. In this example, the real intention of the high-pass filter was to eliminate the large impedance resulting from the resonant condition more than to provide a low-impedance path to all harmonic currents generated at the various VFD sites.

6.10 EXAMPLE 4: COMPARISON BETWEEN SEVERAL HARMONIC MITIGATING SCHEMES USING UNIVERSITY OF TEXAS AT AUSTIN HASIP PROGRAM[4]

This last example describes the results obtained combining a number of filtering schemes using the University of Texas HASIP program. The program carries out harmonic analysis, assuming a fundamental voltage of 1 p.u. and short distances between loads and generators. Zero-sequence harmonics are excluded in the analysis.

The screens shown in Figure 6.16 through Figure 6.24 are self-contained in describing size and characteristics of generation, linear load, non-linear load, capacitor bank, and harmonic filters considered in the analysis. The parameters that were kept fixed are the following:

Generation:
100-MVA base; $Xd'' = 0.20$, $Isc = 500$ MVA
Linear load:
MW: 39
MVAR: 29.3
P.F.: 0.80

FIGURE 6.16 Effect of 5th, 7th, and 11th-harmonic filters, using HASIP.[4]

Harmonic Filtering Techniques

Nonlinear load:
  Six-pulse converter, 25 MVAR, 09 DPF
Harmonic filters:
  Fifth harmonic: 5 MVAR, $X/R = 50$
  Seventh harmonic: 3.6 MVAR, $X/R = 50$
  Eleventh harmonic: 2.2 MVAR, $X/R = 50$

FIGURE 6.17 Effect of 5th harmonic filter, using HASIP.[4]

FIGURE 6.18 Effect of 7th harmonic filter, using HASIP.[4]

126 Harmonics and Power Systems

FIGURE 6.19 Effect of 7th and 11th harmonic filters, using HASIP.[4]

FIGURE 6.20 Effect of 5th and 7th harmonic filters, using HASIP.[4]

Harmonic Filtering Techniques

FIGURE 6.21 Effect of 11th harmonic filter, using HASIP.[4]

FIGURE 6.22 Effect of High Pass filter to provide a low impendance path to harmonics above the 13th, using HASIP.[4]

FIGURE 6.23 Effect of filtering harmonics above the 13th combined with a 50% reduction in generator subtransient impedance, using HASIP.[4]

FIGURE 6.24 Effect of filtering harmonics above the 13th combined with an increase of Distortion Power Factor (DPF) to 1.0, using HASIP.[4]

Harmonic Filtering Techniques

TABLE 6.5
Synopsis of Passive Filter Analysis Using HASIP

NET DPF NET	Q MVAR	Isc/IL	Harmonic Filter of Order				THDI Source	THDV Capacitor	THDI Converter	Figure No.
			5	7	11	>13				
0.91	0.7	7.1	X	X	X	—	2.80%	4.80%	29%	6.16
0.87	0.7	6.8	X	—	—	—	**6.50%**	**9.00%**	29%	6.17
0.87	0.7	6.8	—	X	—	—	**11.80%**	**10.90%**	29%	6.18
0.88	0.7	6.9	—	X	X	—	**13.30%**	**10.30%**	29%	6.19
0.89	0.7	7	X	X	—	—	4.40%	7%	29%	6.20
0.86	0.7	6.7	—	—	X	—	**11.60%**	**10.80%**	29%	6.21
0.87	0.7	6.8	—	—	—	X	**11.30%**	**13.20%**	29%	6.22
0.87	0.7	13.6	—	—	—	X	**12.70%**	**6.80%**	29%	6.23
1	34.7	7.8	—	—	—	X	**10.10%**	5.50%	29%	6.24

Note: Results are plotted in the figures indicated in the last column.

Source: Harmonics Analysis for Ships and Industrial Power Systems (HASIP) Version 1, March 17, 2004, Power Systems Research Group, Department of Electrical & Computer Engineering, The University of Texas at Austin.

Table 6.5 summarizes the different harmonic filtering scenarios tested.

The results indicate the effect of applying different harmonic filtering combinations starting with the application of passive filters of orders 5, 7, and 11. Thereafter, application of individual and different pair sets is tested and, finally, a high-pass filter that reduces harmonic spectral components higher than the 13th harmonic is tested. Figure 6.16 to Figure 6.24 include the waveforms for the capacitor, source current, and nonlinear load current. Also included are the harmonic spectrum of the source and the impedance-frequency characteristics of the system, including filters and capacitor banks. Table 6.5 includes THD values in bold face when they fall in excess of IEEE-519 limits.

Although the results show the expected outcome of the exercise, the inclusion of the interface screen figures illustrates how useful it is to obtain all the information displayed in the process to determine what combination of filters can make harmonic distortion levels fall within recommended limits. It also helps in understanding the role that every element in the network plays in reaching the desired objective.

REFERENCES

1. ANSI/IEEE 519:1992, IEEE Recommended Practices and Requirements for Harmonic Control in Electrical Power Systems, IEEE, 1992.
2. IEEE Std 18–2002, IEEE Standard for Shunt Power Capacitors.

3. The Mathworks, Three-Phase Harmonic Filter, http://www.mathworks.com/access/helpdesk/help/toolbox/physmod/powersys/threephaseharmonicfilter.html.
4. Harmonics Analysis for Ships and Industrial Power Systems (HASIP) Version 1, March 17, 2004, Power Systems Research Group, Department of Electrical & Computer Engineering, The University of Texas at Austin.

7 Other Methods to Decrease Harmonic Distortion Limits

7.1 INTRODUCTION

The first technique to control harmonic-related problems in industry involved substantial use of single-tuned filters to offer a low-impedance path to harmonic currents. Interestingly, it is not difficult to find harmonic-producing loads in the megavolt-ampere range in industry operating with no harmonic filters. This is a difficult issue for power utilities to control because the existing standards are often more of a reference guideline for industry than a regulatory pronouncement. Large harmonic producers, typically in the industrial sector, may be the only producers who adopt harmonic filtering methods to reduce the otherwise multiple disturbances that may arise beyond the metering point and start affecting sensitive equipment and processes. Due to the high cost involved, this is not a common practice in commercial and residential facilities.

Unfiltered harmonic currents are left to spread freely upstream and downstream from the PCC following natural laws of propagation. They may reach adjacent installations and sometimes may even make their way to the utility substation. It is then common to see utilities and harmonic-producing customers in a continuous search for alternative methods to handle and hopefully beat elevated harmonic distortion levels.

For example, special application transformers connected to variable frequency drives and thus highly exposed to harmonic current overheating are usually specified as special K factor transformer designs. These special types of transformer constructions avoid transformer operation below rated values. Type-K transformers are basically designed with improved windings and low loss iron cores that reduce the amount of additional heating produced by harmonic currents. Notice that harmonic currents on the source side of the converter are by no way controlled or eliminated in the windings. Some cancellation of harmonics can take place, for instance, in phase-shift transformers that provide 30° shifting between two six-pulse converters: one fed from delta-connected and the other from wye-connected secondary windings of the transformer.

This chapter describes some of the techniques used in industry to control the flow of harmonic currents produced by nonlinear loads in power systems. The most relevant are the following:

Network reconfiguration
Increase of the short-circuit current ratio
Static multipulse power converters with phase shift transformers
Series reactors
Phase load balancing
Load grouping

7.2 NETWORK TOPOLOGY RECONFIGURATION

One measure often advantageous to reduce the effect of unfiltered harmonics is the reconfiguration of the network. Here it is necessary to identify users and sectors in the installation that introduce large amounts of harmonic currents to the system and to characterize its frequency content. As often occurs in residential installations, redistribution of loads using the same wiring or through additional circuits can provide an economic solution for drastically reducing disturbances.

Distributing the largest nonlinear loads among different feeders, just as balancing single-phase loads in three-phase systems, would be beneficial. This measure would reduce the otherwise excessive voltage drops from harmonic currents carried through a single path.

If harmonic filters are not an option to consider, mixing linear and nonlinear loads on a feeder may allow the reduction of harmonic distortion because linear loads act as natural attenuators of parallel resonant peaks, as discussed in Chapter 8. This measure should not be contemplated when linear loads comprise sensitive electronic or industrial processes, which may be disrupted if THD at some point is somewhat increased.

7.3 INCREASE OF SUPPLY MODE STIFFNESS

The increase of the ratio between the available short-circuit current and rated load current makes a stronger supply node. This happens whenever power utilities increase their substation's size. It also occurs when industrial customers add some cogeneration on the supply bus to help operation during peak demand.

Stiff AC sources increase the available short-circuit current, for which the ratio between short circuit and load currents is often used as a measure of source stiffness. Strong supply nodes can better absorb transient disturbances in the network and attenuate the effects of large transformer inrush currents, cable energization, and start of large motor loads. The same applies for harmonic currents reaching the substation. The reason for this is that the lower impedance of a stiff supply produces smaller voltage drops, not only for steady state but also for higher frequency currents.

High short-circuit currents are then associated with low impedance sources, which are in turn inverse functions of transformer size. This can be illustrated by calculating the change of impedance when an "old" transformer of rating MVA1 is replaced by a "new" transformer rated MVA2. By using the fundamental expression for transformer impedance described in Equation (7.1):

Other Methods to Decrease Harmonic Distortion Limits

$$X_{TRANSF} = \frac{kV^2}{MVA} \times \frac{X_{leakage}}{100}, \quad (7.1)$$

we arrive at the following:

$$\frac{X_{MVA2}}{X_{MVA1}} = \frac{\dfrac{kV_2^2}{MVA_2} \times \dfrac{X_{leakage2}}{100}}{\dfrac{kV_1^2}{MVA_1} \times \dfrac{X_{leakage1}}{100}} \quad (7.2)$$

If we assume all other parameters the same, the impedance ratio in Equation (7.2) reduces to:

$$\frac{X_{MVA2}}{X_{MVA1}} = \frac{\dfrac{1}{MVA_2}}{\dfrac{1}{MVA_1}} = \frac{MVA_1}{MVA_2} \quad (7.3)$$

That is, the impedance ratio of a new to an old transformer varies with the inverse ratio between the old and the new transformers' megavoltamperes. For instance, a 30-MVA transformer would present an impedance twice as small as a 15-MVA transformer of the same voltage class and a two times increase in short-circuit current, assuming the two of them have the same leakage impedance. In other words, the rating of a distribution transformer for a given voltage can be used as an indication of source stiffness.

At harmonic frequencies, inductive and capacitive impedances of the system vary as a function of frequency, as was shown in Chapter 6:

$$X_{Lh} = h\omega_L \quad (7.4)$$

$$X_{Ch} = \frac{1}{(h\omega_C)} \quad (7.5)$$

A stiffer source will primarily affect the inductive component of the system. Harmonic currents will produce voltage drops affected by the inductive reactance of the system, which is composed by feeder and substation components. For the case of short feeders, the source impedance will be the dominant component. In these situations, harmonic currents are likely to reach the substation and the voltage drop on the source impedance, and thus the harmonic distortion will be smaller in stiffer systems.

7.4 HARMONIC CANCELLATION THROUGH USE OF MULTIPULSE CONVERTERS

One-phase converters are used in small-load applications. For lower initial costs, half-wave rectifiers can be applied when current requirements are small. Half-wave rectifying produces a DC component that saturates transformers. To limit the former, the use of full-wave rectifying converters is recommended.

The basic polyphase converter is a six-pulse unit. Theoretically, the 12-pulse unit shown in Figure 7.1(a) will eliminate the lower order harmonics (5th and 7th), for which the first harmonics that will show up are the 11th and the 13th. Because the 17th and 19th are not characteristic harmonics, the following harmonic pair to appear will be the 23rd and the 25th.

Through additional phase multiplication, it is possible to reduce other harmonic currents. For instance, a 24-pulse unit is built up from four 6-pulse rectifier bridges, each of which has a phase shift of 15° relative to the other rectifying units. This is attained by using phase-shifting transformers with separate additional windings connected in zig-zag or in polygon, as illustrated in Figure 7.1(b).

If a six-pulse unit were out of service, some cancellation would still be established with two of the six-pulse units 15° out of phase to one another. However, the third unit would show all of the harmonics typical of a six-pulse converter in the system.

The conditions for eliminating harmonics on a six-pulse rectifier composed of N sections using the phase multiplication approach are the following:

The transformers involved are all of the same transformation ratio and have similar leakage impedances.
The load is split in like parts among the converters.
The firing angle is the same in all converters.
The phase difference between transformers is $60/N$ electrical degrees.

The characteristic harmonics of this harmonic reduction scheme can be expressed as follows:

$$h = kq \pm 1 \qquad (7.6)$$

where
h is the harmonic order.
q is equal to $6 \times N$.
N is the number of six-pulse rectifiers.
k is an integer number (1, 2, 3, …).

If two sections of the rectifier are not equal, noncharacteristic harmonics will always be present as far as the preceding requirements are not met.

Using Equation (7.6), 12-pulse converters would produce cancellation of all harmonics below the 11th and 18-pulse converters would cancel all harmonics below the 17th.

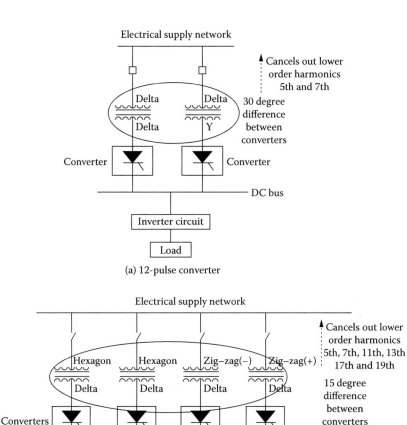

FIGURE 7.1 Phase-shift transformer connections for 12- and 24-pulse converters.

Figure 7.2 describes voltage and current waveforms at load and source locations, along with the harmonic spectrum of the 18-pulse converter.

7.5 SERIES REACTORS AS HARMONIC ATTENUATOR ELEMENTS

Series reactors have been used in industry for a long time as a way to provide some control on short-circuit current levels. We see them in iron and steel or smelting plants and in power substations or neutral-to-ground connection of generators or power transformers. Series reactors are to some extent also used as harmonic

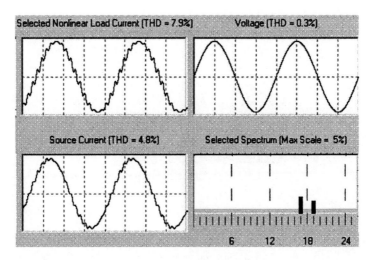

FIGURE 7.2 Eighteen-pulse rectifier.

attenuators in industrial applications. Typically, 5% impedance reactors installed on the source side of power converters are seen in a number of applications.

As an energy-storage device that opposes the rapid variations of current, a series reactor theoretically provides a two-way attenuation to surge and harmonic currents generated on either side of it. This means attenuation of harmonic currents from the converter (or any other nonlinear load) toward the AC source and of harmonic currents from adjacent customers or from surges generated in the distribution system toward the converter. This looks attractive as a way to provide some relief to transient or subtransient types of events on the power line side created during switching of capacitors banks or long cables or transient disturbances created during line faults, in addition to the attenuation of harmonic currents.

7.6 PHASE BALANCING

Some electric power companies use four-wire distribution systems with a primary grounded wye and single-phase transformers supplying phase to ground voltage to single-phase loads such as residential installations, municipal street lighting, etc. Variations in single-phase loads can create unbalanced currents in three-phase conductors, producing dissimilar voltage drops in the three phases and giving rise to phase-to-phase voltage unbalance. Maximum phase-to-phase or phase-to-ground voltage unbalance may be more critical at the far end of a distribution feeder, where voltage may have experienced a substantial drop during heavy load conditions — particularly in the absence of appropriate voltage profile compensation measures.

A perfectly balanced system is difficult to attain because single-phase loads are constantly changing, producing a continuous unbalance of phase voltages and eventually causing the appearance of even and noncharacteristic harmonics.

7.6.1 PHASE VOLTAGE UNBALANCE

The simplest method to determine voltage unbalance is by calculating the greatest deviation of the phase-to-phase voltage from the average voltage as follows:

$$Voltage\_unbalance(\%) = \frac{maximum\ deviation\ from\ average\ voltage}{average\ voltage} \times 100 \quad (7.7)$$

For example, if a 480-V application shows voltages V_{AB}, V_{BC}, and V_{CA} equal to 473, 478, and 486 V, respectively, with an average voltage of $(473 + 478 + 486)/3 = 479$ V, the voltage unbalance is as follows:

$$Voltage\_unbalance(\%) = \frac{7}{479} \times 100 = 1.46\%$$

The amount of voltage unbalance can also be expressed in terms of the negative sequence voltage:

$$Unbalance\_Voltage\_Factor = \frac{Negative\_Sequence\_Voltage}{Positive\_Sequence\_Voltage} \quad (7.8)$$

7.6.2 EFFECTS OF UNBALANCED PHASE VOLTAGE

When the unbalanced phase voltages are applied to three-phase motors, they give rise to additional negative sequence currents that will circulate in the motor windings, increasing heating losses. The most severe condition occurs under an open-phase situation.

All motors are sensitive to unbalance in the phase voltage. Certain kinds of motors, like those used in hermetically built compressors in air-conditioned units, are more susceptible to this condition. These motors operate with elevated current densities in the windings due to the aggregate effect of the refrigerant cooling.

When a motor is suddenly shut down by the protective system, the first step consists in determining the cause of the disconnection and checking the operation current after it has been put back into operation, to make sure that the motor is not overloaded. The next step consists in measuring the voltage in the three phases to determine the amount of voltage unbalance. Figure 7.3 suggests that a motor is undergoing overheating when voltage unbalance exceeds 2 to 3% for a full-load operation.

Computer operation can be affected by a voltage unbalance of 2 to 2.5%. In general, one-phase loads ought not be connected to three-phase circuits that provide power supply to sensitive equipment. A separate circuit should be used for that purpose.

FIGURE 7.3 Derating factor for motors with unbalance in the phase-to-phase voltage. (Adapted from *Optimization of Electric Energy Consumption in Marginal California Oilfields,* EPRI, Palo Alto, CA, California Energy Commission, Sacramento, CA, 2003.)

REFERENCE

1. *Optimization of Electric Energy Consumption in Marginal California Oilfields,* EPRI, Palo Alto, CA, California Energy Commission, Sacramento, CA, 2003.

8 Harmonic Analyses

8.1 INTRODUCTION

Harmonic power system analysis basically requires the same type of information as that required for the analysis of the system under steady-state conditions. The exception to this is the harmonic current source, which must be represented through solid-state switching to recreate the operation of power converters or through appropriate models to represent magnetic core saturation and arc devices. A precise representation of the power system elements will be necessary if an accurate prediction of harmonic response is required.

The propagation of harmonic currents is influenced by a number of factors that relate the offending and affecting parties because the two play a major role in the propagation mode of the harmonic currents in the power system. The waveform distortion produced by a strong harmonic source, for example, may still be tolerable to the power system if its dominant harmonic is farther away from natural resonant points in the system. Conversely, a small harmonic source may give rise to large waveform distortion if any of its characteristic harmonics coincides with a resonant frequency in the system, as was illustrated in Example 3 in Chapter 6.

The growing need to conduct harmonic analyses in electrical power systems makes it convenient to review the fundamental principles that govern the flow of harmonic currents. In the process, we must look at relevant aspects like the importance of linear loads as harmonic distortion attenuation elements. It will be important to find the interaction between the different elements of the circuit in relation to the establishment of parallel resonance points. Ascertaining the relationship between the total harmonic distortion level and the voltage notching caused during the operation of thyristors or any other electronic switching devices in power converters is also important.

This chapter describes the most relevant aspects of the study of waveform distortion caused by harmonic currents propagating from their source of origin to the PCC and spreading further to adjacent locations and even to the remote AC supply (the utility substation). More than a rigorous mathematical procedure to conduct harmonic analysis, the material described presents an overall depiction of factual features, including the simplified approach to the more elaborate models adopted by modern software programs.

8.2 POWER FREQUENCY VS. HARMONIC CURRENT PROPAGATION

It is important to highlight that 60-Hz power flow studies are centered in the steady-state solution of electric networks to establish optimum operating conditions for a given network to satisfy generation and load requirements. A load flow study will investigate system steady-state load performance under normal operating conditions. Power sources including electric company substations and distributed generation are involved. All significant system loads encompassing resistive and inductive elements, capacitor banks associated with power factor, voltage profile, and harmonic filters are involved in load flow analysis.

Source-equivalent models are often assumed, especially in extended networks. Eventually, industrial installations require thorough representations, particularly when detailed characteristics of distributed generation or network topology are desired. A load flow study is usually carried out in power system analysis to determine voltage, current, and power quantities under steady-state operation.

The reasons for conducting a power flow study are diverse:

Determine the flow of active and reactive power required for estimation of power losses.

Assess the requirements of reactive power compensation.

Estimate voltage profile along the feeders, particularly at remote locations, under heavy load conditions. This helps utilities to define corrective actions to compensate sagging voltage profiles along the feeders and maintain voltage within limits stipulated by voltage regulation policies.

Assess loadability limits of distribution systems under different operation scenarios, which can call for the need of resizing conductors and/or transformers. Overloaded feeder sectors and transformers contribute to increased losses.

Harmonic flow studies, in contrast, are conducted to determine the propagation of current components of frequency other than the fundamental and the resultant distortion of the voltage waveform. The aim of these studies, among others, is:

Determine individual and total harmonic distortion levels produced by nonlinear loads at the location of harmonic sources and at the distribution substation.

Determine harmonic resonant frequencies at capacitor bank locations.

Assess the increased losses due to harmonic currents and take action when they approach thresholds that can have an impact on equipment lifetime.

Specify design characteristic of harmonic filters that can permit the reduction of harmonic distortion levels within recommended limits. This is particularly important when severe harmonic distortion produced by certain customer loads penetrates into adjacent customers' installations.

Properly define size of capacitor banks so that the resultant parallel peak impedance stays away from characteristic harmonics of harmonic-producing nonlinear loads.

Figure 8.1 shows voltage and current waveforms for cases when (a) voltage is in phase with current; (b) voltage leads current; and (c) current leads voltage involving a power factor equal to 0.7. These results are obtained in a typical load flow

Harmonic Analyses

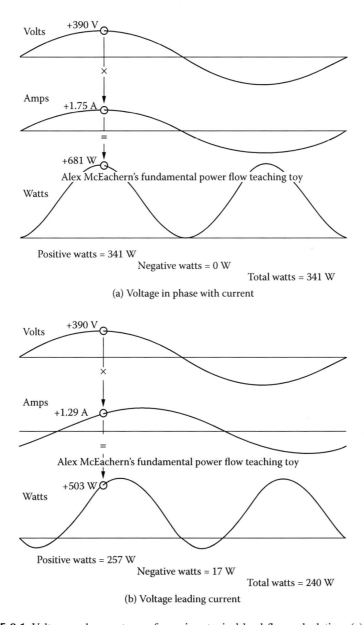

FIGURE 8.1 Voltage and current waveforms in a typical load flow calculation. (a) Voltage in phase with current; (b) voltage leading current; (c) current leading voltage.

Continued.

FIGURE 8.1 *Continued.*

calculation. Figure 8.2(a) through Figure 8.2(c) depict voltage and current graphs involving third, fifth, and seventh harmonic currents, respectively. The lower graph on each plot is the active power. "Positive watts" refer to power delivered from the source to the load. "Negative watts" refer to power returned from the load storage devices (inductive and capacitive elements) to the source. Notice that the total active power is equal to zero for the three cases in Figure 8.2, which is valid for zero phase angle between voltage and current waveforms. These plots were generated with the free harmonic tool from Power Standards Lab.[1]

8.3 HARMONIC SOURCE REPRESENTATION

8.3.1 Time/Frequency Characteristic of the Disturbance

The effects of harmonic sources on the power system will always be more appropriately assessed in the frequency domain — i.e., through a comprehensive Fourier analysis of the system. This requires using manufacturer or measured data to represent harmonic sources from every existing and future nonlinear load in the simulation study. Harmonic current spectra of different harmonic generating equipment or appliances usually include magnitude and phase angle.

The representation of harmonics as ideal current sources assumes that voltages are not distorted. For some nonlinear devices, the representation is considered accurate as long as the real voltage distortion is below around 10%.[2]

Harmonic modeling techniques involve the representation of distortion-producing loads in a form in which they can realistically represent the harmonic sources in the power system network. Harmonic spectra of the load current (see Chapter 2)

Harmonic Analyses

(a) Voltage and current with 3rd harmonic

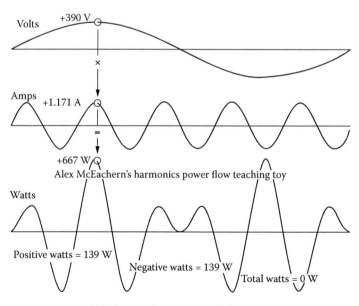

(b) Voltage and current with 5th harmonic

FIGURE 8.2 Voltage and current waveforms containing harmonic currents. (a) voltage and current with third harmonic; (b) voltage and current with fifth harmonic; (c) voltage and current with seventh harmonic. *Continued.*

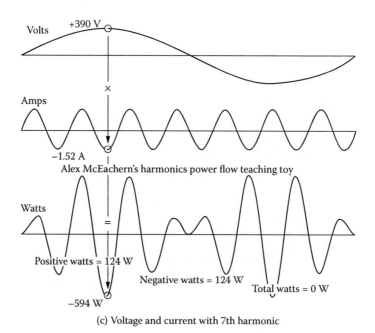

(c) Voltage and current with 7th harmonic

FIGURE 8.2 *Continued.*

describe spectral components of individual harmonic sources. In harmonic analysis, these current sources are injected on the electrical system at the point at which they are created, i.e., at the location of the nonlinear load. This is equivalent to superimposing the harmonic currents on the load current waveform.

As described in Chapter 2, there can be multiple sources of harmonic distortion and every one of them may include different harmonic components. Therefore, harmonic current injection techniques will generally entail the representation of a number of different spectral components that usually fall under one of the following categories:

Power electronic devices
Arc-type devices (electric furnaces, fluorescent lamps, etc.)
Ferromagnetic devices (transformers, induction motors, etc.)

For solid-state devices, characteristic harmonic content can be determined in a straightforward way based on the number of the rectifier steps. For the cases involving arc devices and transformers, it is necessary to obtain the typical harmonic spectra.

Figure 8.3 and Figure 8.4 show typical spectra of 6- and 12-pulse rectifiers in power converters, respectively. They represent the order and amplitude of every harmonic current that will be injected at the location of the nonlinear load. Together with the harmonic spectrum, the nonlinear load current, source current, and power

FIGURE 8.3 Six-pulse rectifier.

FIGURE 8.4 Twelve-pulse rectifier.

factor correction capacitor voltage waveform are shown. This example was obtained using the unrestricted University of Texas Harmonic Analysis Program (HASIP).[3]

The decomposition of a signal in its spectral components, as described in Chapter 1, can be carried out using the following formulas:

$$f(t) = \frac{a_0}{2} + \sum_{0}^{\infty}(a_n \cos n\omega t + b_n \sin n\omega t) \tag{8.1}$$

where $\omega = 2\pi/T$ and n is the harmonic number, with

$$a_n = \frac{2}{T}\int_0^T f(t)\cos nwt.dt \qquad (8.2)$$

for $n \geq 0$ and:

$$b_n = \frac{2}{T}\int_0^T f(t)\sin nwt.d \qquad (8.3)$$

for $n \geq 1$.

Harmonic sources are generally dispersed and are usually modeled as current sources of a frequency corresponding to desired harmonic current. Figure 8.5 portrays a typical model of harmonic current simulation involving a number of sources.

Most software tools include typical harmonic sources so that the user does not need to exercise additional efforts in building them to make an assessment of the problem, particularly during planning stages of a network. In some instances in which harmonic resonance is suspected to be the source of specific disturbances, harmonic current spectra from measurements are preferred. These will provide a far more accurate representation of the harmonic source.

The degree of voltage signal distortion will depend on the amplitude of the harmonic current source and on its propagation on the network. Therefore, an industrial customer may produce harmonic currents that can create some degree of distortion on the voltage waveform at adjacent customer locations. The level of

XC1, XC2 = Reactance of power factor correction capacitors

XL1 – XL4 = Step down transformers

XS = Source impedance

FIGURE 8.5 Harmonic injection current from different sources.

Harmonic Analyses

FIGURE 8.6 Impedance vs. frequency response of a typical distribution system.

distortion will depend on how much harmonic current will flow toward the source and how much of it will be shared with adjacent facilities.

Inductive and capacitive impedances play an important role in the harmonic current propagation phenomenon. Connecting service drops, transformers, and capacitor banks are some of the elements that can contribute to harmonic current damping or to the excitation of resonant frequencies that can produce significant amplification of voltage distortion.

Impedance scans are used to produce an overall representation of the system response as a function of frequency at specific network locations. This impedance vs. frequency characteristic is generally determined at locations where nonlinear loads, capacitor banks, or harmonic filters exist in the network. They can pinpoint capacitor bank resonant conditions like that occurring near the seventh harmonic in Figure 8.6 and they can depict the system response when a fifth harmonic filter is applied, as demonstrated in the same example.

Impedance scans are an excellent tool to anticipate system response in planning network or load expansions.

8.3.2 Resonant Conditions

Natural resonant conditions are important factors that affect the harmonic levels and total harmonic distortion of the system. Harmonic parallel resonance is a large impedance to the flux of harmonic currents; series resonance is a small impedance only limited by a resistive element. When resonant conditions are not a problem, it is because the system has the capacity to absorb important amounts of harmonic currents.

Series harmonic resonance is the result of the series combination of two elements in the electric network seen from the nonlinear load. For instance, a power factor capacitor bank at the primary side of the service transformer feeding a nonlinear load and the inductance of the transformer form a series LC circuit seen from the harmonic current source, as illustrated in Figure 8.7(a). This is actually the operation principle of a single-tuned harmonic filter, in which the series reactor, more than being provided by the transformer reactance, is physically a separate reactor in series with the filter capacitor, as described in Chapter 6.

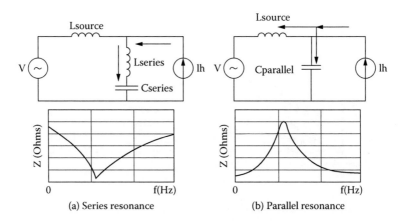

FIGURE 8.7 Series and parallel resonance and its Z–f plot.

However, there are instances of series resonance in which unintended harmonic filters are formed at customer installations involving power factor capacitors; these should be further assessed. Equation (8.5) in Section 8.7 can be used to determine the risk for capacitor bank damage if a resonant condition relating to a characteristic harmonic of the load occurs.

Series harmonic resonance presents a low impedance trajectory to the harmonic currents. The resultant large currents may produce telephone interference in nearby telecommunication systems, conductor heating, and excessive voltage distortion on capacitor banks.

Parallel harmonic resonance (see Figure 8.7b) occurs when the reactance of an inductive and a capacitive element, which from the harmonic source appear like parallel elements, become equal at a given frequency. If this frequency coincides with or falls near one of the characteristic harmonics of the load, this harmonic current will oscillate in the form of trapped energy between the inductive and the capacitive elements of the circuit. To the harmonic current, this is equivalent to a large upstream impedance and, as a consequence, a large voltage drop at the frequency of the harmonic current will take place. Overvoltage conditions that can exceed safe operating limits in capacitor bank may develop.

Because inductance and capacitance are intrinsic elements in a power system, series and parallel resonance phenomena are occurrences that, to some extent, are always present in distribution systems.

8.3.3 Burst-Type Harmonic Representation

There are phenomena in power systems operation that produce waveform distortion for short periods. Line-to-ground faults, inrush currents during transformer energization, kick-in of large loads, and motor starting can be mentioned as typical examples of short-term (burst) harmonic distortion. The modeling of these phenomena is similar to the modeling of steady-state harmonic waveform distortion, except

Harmonic Analyses 149

that Fourier transform analysis must perform averaging over time windows consistent with the duration of the disturbance.

8.4 HARMONIC PROPAGATION FACTS

In general, harmonic currents propagating on a power system follow the same physical laws that govern the propagation of low-frequency phenomena. Some relevant factors that play a role in this process are the following:

- Location of the injecting harmonic source on the network:
  - At an individual customer facility
  - At the location of a large harmonic producer at an industrial facility at which the potential for harmonic currents propagating upstream and reaching adjacent customer installations is found
  - As sources from multiple customers interacting with one another
- Topology of the power system sourcing the harmonic-producing load. As a simple rule, customers fed off from a feeder in the near proximity of the substation are expected to experience harmonic distortion levels somewhat smaller than those of customers fed off downstream from the far end of the feeder. The reason for this is the smaller source impedance, which, in the case of far end customers, is increased by the per-unit length impedance of the feeder.
- In connection with the previous point, harmonic source stiffness is important in defining the extent of waveform distortion. Weak systems are associated with a large source impedance and stiff systems are associated with a small impedance. Therefore, weaker systems will produce larger voltage drops from harmonic currents than stiff systems. This will have an impact on the total harmonic waveform distortion.
- Power system capacitance can give rise to series and parallel resonant conditions that can magnify harmonic currents and voltages, for which it should always be considered in harmonic assessment studies.
- Distribution system capacitance includes capacitor banks used for power factor correction and for control of voltage profile. Large sections of insulated cables in underground commercial and residential areas can also introduce significant capacitance.
- Transmission line capacitance is generally neglected in harmonic propagation studies because the representation of the network rarely extends beyond the distribution substation. However, it may be included in cases in which more than one large industrial facility is fed off from the same transmission line.
- Linear load effects are important in defining the extent of penetration of harmonic currents into the electric network. Minimum "linear" loading leads to small attenuation of harmonic distortion and can be used to establish the maximum levels of harmonic distortion that can be expected. Likewise, maximum "linear" loading means greater attenuation of harmonic

distortion and can reveal the degree of minimum level of harmonic distortion that should be expected.

8.5 FLUX OF HARMONIC CURRENTS

How do harmonic currents flow in an electrical network? The easy answer to this question is that harmonic currents freely flow from their source of generation to whatever point electrically connected to them. However, harmonic currents will find it easier to propagate on reduced impedance paths. What this means is that, once we apply Fourier decomposition techniques to find the harmonic components of the current, we can treat every single harmonic source as we would a 60-Hz current.

The essential difference with power frequency currents, however, is that network impedances are frequency dependent, as described by Equation (7.4) and Equation (7.5) of Chapter 7. The consequence of this is that higher order harmonics will see a large inductive reactance and a small capacitive reactance. Conversely, lower order harmonics will propagate through small inductive and large capacitive paths.

This simple fact allows us to understand that higher order harmonic currents produced at the end of an uncompensated feeder will not have the same penetration towards the distribution substation that lower order harmonics will. Using the resonant frequency equation,

$$f_{res} = \frac{1}{2\pi\sqrt{LC}} \qquad (8.4)$$

we can also realize that the same capacitor bank would present a lower resonant frequency when interacting with a large inductive component, e.g., if it were located at the end of a feeder rather than if it were located close to the substation.

We would then find all possible combinations of L and C, but maybe one that is relevant is when L and C are large, which would combine to produce a resonant point on the lower region of the frequency spectrum where the largest characteristic harmonic currents of six-pulse converters, fluorescent lighting, and electric furnaces are found.

Harmonic current propagation thus depends on the electrical parameters of the circuit in the same way as a fundamental frequency power flow starting from the point where it is created (the location of nonlinear loads) and spreading upstream throughout utility source impedance elements, which is generally small. However, harmonic currents may also reach adjacent customer locations where power factor correction capacitors offer low impedance paths at high frequencies. See Figure 8.8.

In the absence of harmonic filters, harmonic currents not only can reach the AC source and produce undesirable harmonic distortion levels but also may produce transformer overheating and other unfavorable effects discussed in Chapter 4.

The penetration of harmonic currents into the electrical network has become a hot issue for electric utilities in the light of stringent regulations calling for improved power quality indices and compliance with industry standards. IEEE-519[3] on

FIGURE 8.8 Harmonic currents and their propagation to distribution network and adjacent installations.

FIGURE 8.9 Harmonic current flow on a typical distribution network.

harmonic recommended limits is undergoing a working group revision aimed at producing specific electric utility application guidelines.[4,5]

The main concern with harmonic distortion is the steady increase of nonlinear loads at commercial and industrial facilities, including all kinds of power converters from massive personal computer utilization, among others. Figure 8.9 depicts a simplified analysis model for the one-line diagram of Figure 8.8, showing line and transformer impedances. The aim of this illustration is to show how a large harmonic current injected by a particular customer has the potential to penetrate through multiple paths into the distribution network. It can reach adjacent customer installations and can even propagate to the AC power source, where it can create an undesirable increase of harmonic distortion. Notice that the flow of harmonic currents on that part of the feeder between harmonic and AC sources is opposite to the normal flow of power frequency current.

8.5.1 Modeling Philosophy

The AC source is usually represented through a simple Thevenin equivalent using the short circuit impedance. Some computer modeling programs offer very detailed representation of AC sources, including excitation system in synchronous generators, mechanical, and electrical constants. This provides the possibility to incorporate in-depth representations of additional sources of energy in the simulation model. These may include distributed generation used by utilities to cope with voltage and power support during peak demand hours or cogeneration resources in industry.

Transmission lines should be modeled when a number of large industrial customers are fed off from the same transmission line because, due to their lower impedance, harmonic currents are little attenuated and may easily spread to adjacent facilities.

Distribution feeders are represented by their per-unit-length positive sequence impedance, usually comprising resistance and inductance for overhead lines and including capacitance for underground feeders. Harmonic analysis software usually offers a number of power line models ranging from constant frequency (Bergeron) to frequency-dependent models. In all cases, conductor configuration and soil parameters are needed so that the program can calculate the necessary impedance/admittance matrices.

For industrial loads fed off from dedicated feeders, it is important to set up a detailed representation of the low-voltage components, including linear and nonlinear loads as well as any kind of energy storage devices like capacitor banks and reactors. On the primary side of the service transformer, a high-voltage side equivalent impedance can be used as a first approach because the transformer is the dominant impedance.

Capacitor banks should be modeled at the primary and secondary sides of the transformer whenever present because they are important elements when resonant conditions are excited. At the primary side, the capacitance of the distribution line is not significant. However, insulated cables of considerable length should be modeled because, as capacitive elements, they may play an important role in establishing the level of telephone interference, generally produced by higher order harmonics.

8.5.2 Single-Phase vs. Three-Phase Modeling

For most harmonic studies, the three-phase model representation of the system using the positive sequence parameters will be sufficient. Exceptions to this statement may be found in the following cases:

> Unbalanced systems. In this case, the unbalanced system or the uneven phase harmonic sources can be incorporated to determine the per-phase harmonic components precisely. If telephone interference is to be assessed, higher order triplen harmonics (produced, for instance, during transformer saturation) will be of special interest because they are added in phase and can be present at considerable distances from their source.
>
> One-phase capacitor banks. Although rarely found, single-phase compensation on three-phase systems will also require per-phase models.

For one-phase models, the per-phase representation is carried out in separate form for the different propagation modes and thereafter converted to per-phase quantities. This is done through eigenvector analysis.

8.5.3 Line and Cable Models

Most software tools include some form of fixed or frequency-dependent models for transmission lines as described previously. Frequency-dependent quantities are the

Harmonic Analyses

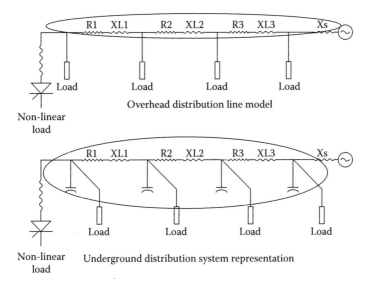

FIGURE 8.10 Model representation of overhead and underground distribution systems.

earth return, which is calculated from Carson's equations,[6] and skin effect, which is generally derived using Bessel functions.[7] For short lines and/or low-frequency applications, a series impedance may provide a reasonably good representation of the line. However, for underground cables or lines with significant portions of insulated cables or in studies that involve higher order harmonics, it will be important to include the shunt capacitance in their representation, as depicted in Figure 8.10. For multiphase lines, equations are solved using modal transformation as described in reference 8.

For transmission lines, the representation of long lines must consider transpositions and distributed capacitances for a more adequate modeling.

8.5.4 Transformer Model for Harmonic Analysis

Transformers have two components that are of most interest:

Leakage impedance
Magnetizing impedance

Normally, the magnetizing impedance is much larger than the leakage impedance when the transformer in not operated on the saturation region. For harmonic analysis, a representation of the transformer should include a current source[9] and a resistance (Rm) to account for core losses, as depicted in Figure 8.11. For low-order harmonics, a lumped impedance representation for the leakage impedance can be used.

When required by the model used, the X/R ratio can be assumed to have a typical value of ten and two or three times as high for transmission transformers. If the transformer is not a significant source of harmonics, the magnetizing impedance can be ignored.

FIGURE 8.11 Transformer model for harmonic analysis. (Adapted from *IEEE Trans. Power Delivery,* 11(1), January 1996, 452–465.)

8.5.5 Power Factor Correction Capacitors

Capacitor bank placement in distribution or industrial networks will influence harmonic resonance frequency. Another consequence of applying capacitor banks in installations with nonlinear loads is that the total power factor will not be influenced in the same way as the displacement power factor. Real power factor will be smaller than displacement power factor, as described by Equation (1.47) and Equation (1.48) in Chapter 1.

Applying capacitor banks to nonlinear loads can have the effect of increasing waveform distortion on voltage and current signals at the location where capacitor banks are applied as well as at adjacent installations, for which in some specific cases it may be important to extend the analysis to neighboring facilities to get a complete picture of harmonic penetration effects. Single-phase capacitor banks in unbalanced systems may produce an increase in noncharacteristic harmonics.

8.6 INTERRELATION BETWEEN AC SYSTEM AND LOAD PARAMETERS

The following elements of the electric power system are important to consider in the study of harmonic propagation:

Step-down transformers
Resistive components of the load
Rotating machine components

Step-down transformers are important at high frequencies because they form a series reactance with the load. System elements that absorb active power, such as resistive and inductive components, can become significant attenuators of harmonic waveform distortion.

Step-down transformers are usually represented by means of their series leakage reactance. Magnetizing current of the transformer is typically a small percentage

Harmonic Analyses

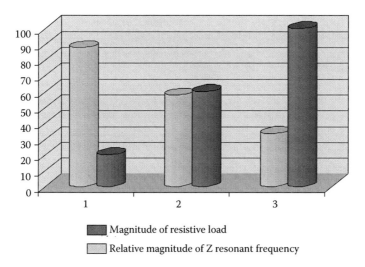

FIGURE 8.12 Typical resistive load effects on the parallel resonant peak impedance.

(normally below 3%) of its rated load for which the magnetizing branch is often omitted in analysis. However, most software tools offer suitable models to include transformer saturation curves when needed. This could be of interest, for example, when investigating zero sequence currents in delta-/Y-connected transformers.

Linear loads are generally represented by passive elements. At the frequency of the lower order harmonics, transformer leakage reactance is small compared with the resistive impedance of the load. However, at the frequency of the higher order harmonics, the step-down transformer reactance can become as large as that of the load, so it can be considered that the transformer represents an important attenuator of high-frequency harmonic currents generated by the nonlinear load, which is favorable to the source.

The resistive portion of the load provides damping that affects system impedance in the proximity of the resonant point. Therefore, the higher the loading of the circuit is, the lower the resulting impedance near the resonant frequency is. The trajectory of the resistance offers a smaller impedance at higher load levels. This is the path taken by harmonics when parallel resonance arises. Typical distribution system response with different load levels is shown in Figure 8.12 at the frequency corresponding to the system's parallel resonance.

Electrical motors can be represented by their short-circuit or blocked rotor impedance at harmonic frequencies. The subtransient impedance is more common to represent induction motors. Often an average of the direct and quadrature axes' impedances is used. This impedance does not provide a significant attenuation of the parallel resonant peak. It can, however, slightly change the resonant frequency. Likewise, load variations can also produce a similar resonant frequency shift.

Balanced systems are modeled using their positive sequence impedances. This includes loads and reactive compensation elements. Under these conditions, harmonic currents will have characteristic phase sequences, as discussed earlier in the section dealing with harmonic sources.

For representation of unbalanced systems comprising uneven distribution of loads on the three phases, single-phase capacitor banks, or asymmetrical configurations of power cables, a per-phase representation using positive, negative, and zero sequence impedances should be used. Results of the analysis may reveal even harmonics.

Large harmonic distortion should be expected when nonlinear loads account for the greatest percentage of the loads in a given application. This can be a case of large VFDs fed by dedicated feeders as the only load. Some examples include offshore production platforms in the oil industry and steel mill plants.

Some form of parallel resonance should be suspected when a large harmonic voltage is detected at the source terminals under light load conditions involving a nonlinear load application.

8.6.1 Particulars of Distribution Systems

There might be a variety of types of nonlinear loads. Frequency spectra can usually be obtained from manufacturers. Harmonic analysis software usually includes some generic models for 6-, 12-, 18-, and 24-pulse converters.

Cogeneration is sometimes used in industrial plants such as steel mills and others. Because adding harmonic filters in variable frequency drive installations is likely to convert power factor from lagging to leading, it will be important additionally to check for generator stability. A representation of the machines, including machine constants, is necessary if transient stability or machine response to any other transient condition is assessed.

Nonlinear loads in industry are often cyclic, for which the characterization of harmonic distortion levels may require carrying out long-term measurements to determine the profile of the voltage and current waveform distortion. This is important when harmonic filtering is being considered so that filter components are properly sized to match the maximum load variations.

If we are concerned with harmonic currents making their way to the distribution substation, it will be important to consider that the most common harmonic currents observed at a distribution substation are the fifth and the seventh. Existing capacitor banks may be tuned to those frequencies by using series reactors if the resonant frequency is found around those frequencies.

Utilities may find it more convenient to consider applying a number of voltage regulation capacitor banks along distribution feeders instead of a large capacitor at the end of the feeder. This will avoid producing a single parallel resonant point on the lower frequency spectrum where characteristic harmonics of the load are most likely to be encountered and induce the creation of multiple higher frequency resonant points at less troublesome locations on the spectrum.

This is easier to understand if, on the resonant frequency equation,

$$f_{res} = \frac{1}{2\pi\sqrt{LC}},$$

we keep C constant and let L increasingly change.

8.6.2 SOME SPECIFICS OF INDUSTRIAL INSTALLATIONS

As illustrated in Figure 8.13, industrial power systems look like compact distribution systems. Some of them may even have a generation facility running continuously or during peak demand periods to help the power utility cope with voltage regulation problems. However, some important differences must be mentioned:

- Capacitor banks often dominate the frequency response of industrial systems and the short-circuit inductance is relatively large.
- The parallel resonance is often found at low-order harmonics.
- The amount of harmonic producing loads in industrial systems is often greater than in distribution systems. This is due to the use of large power converters, arc furnaces, variable frequency drives, etc.
- Resistive loads are often small and resonant impedance peaks are not optimally damped near resonant frequencies. This often results in severe harmonic distortion, although, to a lesser extent, motor loading does help to provide some attenuation of the resonant peak impedance peak.
- Most industrial installations can be modeled as balanced systems. Loads are generally composed of three phases in nature and are balanced, including harmonic sources and transformers, and capacitor banks are three phased and balanced.

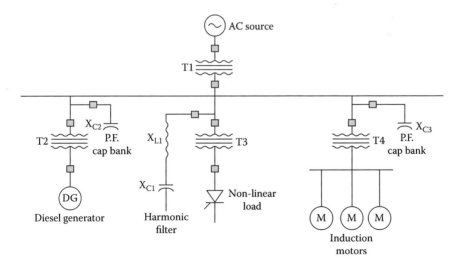

FIGURE 8.13 Typical industrial power system.

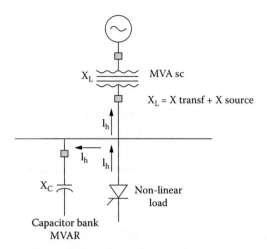

FIGURE 8.14 Circuit for simplified calculations.

8.7 ANALYSIS METHODS

8.7.1 Simplified Calculations

An oversimplified manual calculation suggested in reference 3 may be used in cases when the system can be represented by the circuit of Figure 8.14. Among the most important calculations for this circuit, we have the system resonant frequency, which is obtained with the following expression:

$$Sys\_Res\_freq = \sqrt{\frac{Short\_circuit\_MVA}{Cap\_Bank\_MVAR}} = \sqrt{\frac{Cap\_Bank\_X_C}{Substation\_Short\_circuit\_X_L}}$$

$$= \sqrt{\frac{Transformer\ kVAx100}{Rating\ of\ connected\ capacitors\ in\ kVA\ X\ Transformer\ Z\%}} \quad (8.5)$$

where:

Sys_Res_freq is the system resonant frequency expressed as an integer multiple of the fundamental frequency.

Short_circuit_MVA is the system short-circuit megavoltamperes at the point under consideration.

Cap_Bank_MVAR is the rated megavoltamperes of the capacitor bank at the system voltage.

Cap_Bank_X_C is the reactance of the capacitive bank at the fundamental frequency in ohms.

Substation_Short_circuit_X_L is the short-circuit reactance at the supply substation in ohms.

Harmonic Analyses

If the calculated resonance frequency is close to one of the characteristic harmonics of the source (the nonlinear load), large harmonic overvoltages that may increase waveform distortion to inconvenient levels can develop. A precise and more rigorous study is then required.

In a simplified manner, the system impedance expressed in the frequency domain can be approximated by:

$$Z(\omega) = \frac{(R + j\omega L)}{[1 - \omega^2(LC) + j\omega RC]} \quad (8.6)$$

where $Z(\omega)$ = system impedance as a function of frequency ($\omega = 2\pi f$).

The harmonic voltage at every individual harmonic frequency can then be obtained using:

$$V_h = (I_h)(Z_h) \quad (8.7)$$

where I_h is the current source at the characteristic harmonic h.

Taking into account the contribution of every harmonic considered in the assessment, the RMS voltage can be determined from:

$$V_{RMS} = \sqrt{V_1^2 + \sum_{h=2}^{n} V_h^2} \quad (8.8)$$

Total harmonic distortion (THD) and the telephone influence factor (TIF) can also be estimated through a simplified approach using Equation (1.39) and Equation (1.42) in Chapter 1. More rigorous calculations can be carried out using individual harmonic voltages and currents.

8.7.2 Simulation with Commercial Software

When circuits become complex, it is necessary to conduct computer simulations, using software capable of carrying out the following calculations:

The frequency response of the system
The frequency response for multiple harmonic sources
The solutions for unbalanced polyphase circuits

The most common method employed in harmonic analysis programs is to make a direct solution of the matrix impedance at multiple frequencies. With this type of solution, nonlinear devices are modeled as ideal current sources at harmonic frequencies. Frequency dependence of the system elements (transmission lines, transformers, motors, etc.) is generally included in calculations. However, the system is considered linear at every individual frequency.

FIGURE 8.15 Circuit diagram of the model used to study transformer inrush current.

The iterative Newton–Raphson method is commonly used for the solution. The implementation of this solution is done for balanced systems and is generally applicable to the analysis of transmission systems and in distribution networks.

8.8 EXAMPLES OF HARMONIC ANALYSIS

To illustrate the versatility of modern computer simulation software, a couple of cases are presented to illustrate the harmonic analysis on the waveform distortion produced during transformer energization (inrush current) and during a single-phase fault to ground. These cases are chosen to complement other harmonic analysis cases shown throughout the book.

8.8.1 HARMONIC CURRENT DURING TRANSFORMER ENERGIZATION

This example is taken from the PSCAD User's Group home page.[10] It calculates the inrush current on closing the transformer breaker near the zero crossing of the breaker. Figure 8.15 shows the electric diagram of the source and the transformer. Transformer data are provided in Table 8.1.

Figure 8.16 shows plots for the current, voltage, and flux of the transformer when the transformer breaker is closed near the zero crossing of the voltage. Using the Electrotek Concepts TOP Output processor,[11] FFT analysis is applied to the transformer inrush current and shown in Figure 8.17. Here we can see that the dominant 60-Hz harmonics are the 2, 3, and 5 and the calculated total harmonic distortion reaches a level of 103%.

8.8.2 PHASE A TO GROUND FAULT

This example considers a single-phase-to-ground fault at a location F2 (between circuit breaker B1 and transmission line in Figure 8.18). Here, again, this is not the type of steady-state harmonics found in the operation of a power converter, but serves well to describe the voltage and current waveforms and their spectral content during a fault to ground. Protective engineers need to consider the distortion imposed on the current waveform during short-circuit conditions for adequate relay setting of protective devices. The PSCAD model in Figure 8.18 includes a long transmission line linking two substations and the location of the fault at point F2 downstream of circuit breaker B1.

TABLE 8.1
Transformer Data Including Saturation Curve Used to Test Inrush Current

Transformer name	T1	Magnetizing current at rated voltage			
Transformer MVA	1 [MVA]	Enable saturation		1	
Primary Voltage (rms)	1 [kV]	Point 1 – Current as a %of rated current		0.0 [%]	
Secondary Voltage (rms)	1 [kV]	Point 1 – Voltage in p.u.		0.0 [p.u]	
Base Operation Frequency	60 [Hz]	Point 2 – (I-V)	0.1774 [%]	0.324129 [p.u.]	
Leakage Reactance	0.010 [p.u.]	Point 3 – (I-V)	0.487637 [%]	0.61284 [p.u.]	
No load losses	0.0 [p.u.]	Point 4 – (I-V)	0.980856 [%]	0.825118 [p.u.]	
Copper losses	0.0 [p.u.]	Point 5 – (I-V)	2 [%]	1.0 [p.u.]	
Model saturation?	Yes	Point 6 – (I-V)	3.09543 [%]	1.08024 [p.u.]	
Tap changer winding	None	Point 7 – (I-V)	6.52348 [%]	1.17334 [%]	
Graphics display	Windings	Point 8 – (I-V)	20.357 [%]	1.26115 [p.u.]	
		Point 9 – (I-V)	60.215 [%]	1.36094 [p.u.]	
		Point 10 – (I-V)	124.388 [%]	1.49469 [p.u.]	

FIGURE 8.16 Current, voltage, and flux waveforms during transformer energization.

FIGURE 8.17 Inrush current harmonic spectrum.

Figure 8.19 and Figure 8.20 show the voltage and current on every phase at the time of the fault at the locations of breakers B1 and B2, respectively.

Finally, Figure 8.21 depicts the harmonic spectra of voltage and current harmonic waveforms at the locations of breakers B1 and B2 after B1 and B2 have reclosed. The calculated THD levels are around 10 and 45% at breaker B1 and around 10 and 43% at breaker B2, for voltage and current, respectively.

Harmonic Analyses

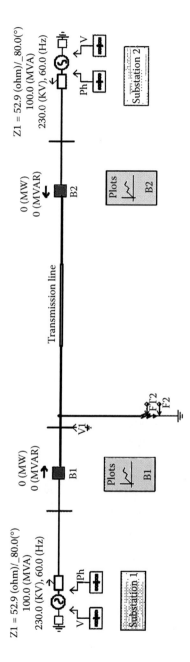

FIGURE 8.18 Model used to calculate harmonic content during a fault to ground.

FIGURE 8.19 Voltage and current waveforms during the simulated phase A to ground fault at location B1.

Harmonic Analyses

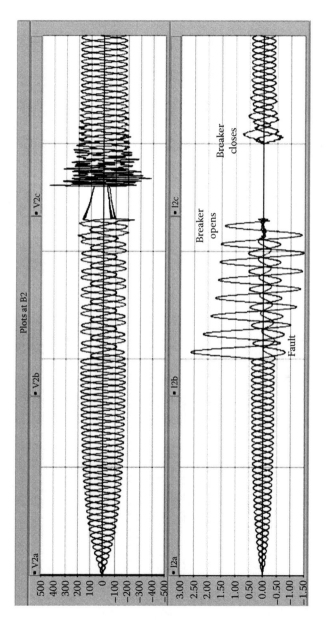

FIGURE 8.20 Voltage and current waveforms during the simulated phase A to ground fault at location B2.

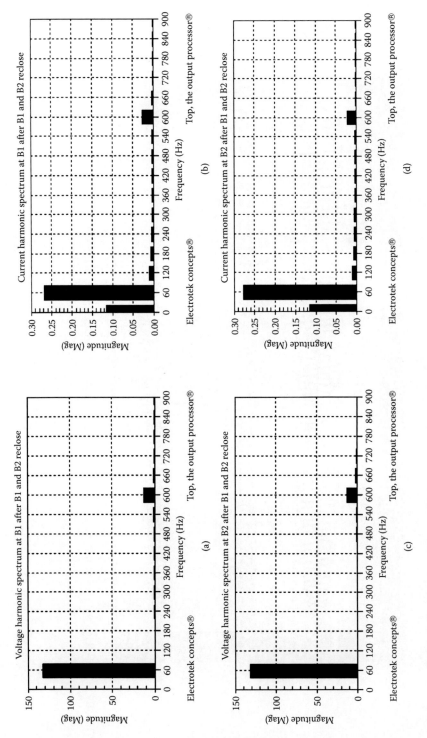

FIGURE 8.21 Voltage and current spectra at locations B1 and B2 after the two breakers have reclosed.

REFERENCES

1. McEachern, A., Power Quality Teaching Toy 2.0.2, Power Standards Lab website: http://www.PowerStandards.com.
2. IEEE-519:1992, Recommended Practices and Requirements for Harmonic Control in Electric Power Systems.
3. Harmonics Analysis for Ships and Industrial Power Systems (HASIP) Version 1, March 17, 2004, Power Systems Research Group, Department of Electrical & Computer Engineering, The University of Texas at Austin.
4. Halpin, M., Comparison of IEEE and IEC harmonic standards, *Proc. 2005 IEEE Power Eng. Soc. Gen. Meet.*, June 12–16, 2005, San Francisco, CA.
5. Halpin, M., Harmonic modeling and simulation considerations for interharmonic limits in the revised IEEE standard 519:1992, *Proc. 2005 IEEE Power Eng. Soc. Gen. Meet.*, June 12–16, 2005, San Francisco, CA.
6. Galloway, R.H. et al., Calculation of electrical parameters for short and long polyphase transmission lines, *Proc. IEEE*, 111(12), Dec. 1964, 2051–2059.
7. Magnunson, P.C., *Transmission Lines and Wave Propagation*, Allyn and Bacon, Boston, 1965.
8. Dommel, H.W., *Electromagnetic Transients Program Reference Manual* (EMTP Theory Book), prepared for Bonneville Power Administration, Department of Electrical Engineering, University of British Columbia, Aug. 1986.
9. Bonner, A., Grebe, T., Gunther, E., Hopkins, L., Marz, M.B., Mahseredjian, Miller, N.W., Ortmeyer, T.H., Rajagopalan, V., Ranade, S.J., Ribeiro, P.F., Shperling, B.R., and Xu, W., TF on harmonic modeling and simulation, modeling and simulation of the propagation of harmonics in electric power systems, *IEEE Trans. Power Delivery*, 11(1), January 1996, 452–465.
10. PSCAD/EMTDC User's Group home page, Transformer inrush current simulation, http://pscad_µg.ee.umanitoba.ca/index.htm.
11. Electrotek Concepts, TOP — The Output Processor, http://www.pqsoft.com/top/.

9 Fundamentals of Power Losses in Harmonic Environments

9.1 INTRODUCTION

Estimation of harmonic-related losses in distribution systems entails the knowledge of harmonic sources, the characteristics of the elements involved in the propagation of the harmonic currents, and — what is most important and probably the most difficult to assess — the period during which harmonic currents are present in the system. Some applications involve well-identified periods of operation — for instance, the use of fluorescent lighting in commercial installations or the operation of electronic and digital equipment in business and other commercial facilities during working hours. Industrial facilities, however, are a special case because a variety of automated processes take place, many of them cyclic and often involving a mix of linear and nonlinear loads.

On the other hand, the operation of the electrical system, including operating voltage, substation and service transformer configurations (primary and secondary connection types), and their leakage impedance, voltage regulation, and reactive power management practices, also play an important role in the estimation of losses.

The utilization of harmonic cancellation schemes that eliminate higher order harmonics is also relevant because losses are associated to the square of the current. Finally, the mobility of parallel resonant points as capacitor banks are turned on and off is also key for the estimation of harmonic related losses.

This chapter presents a general description of the most relevant aspects to bear in mind when looking at losses in electrical systems related to harmonic currents.

9.2 MEANING OF HARMONIC-RELATED LOSSES

Increased rms values of current due to harmonic waveform distortion leads to increased heat dissipation in equipment and undesired fuse operation in capacitor banks. The resulting effect can affect life cycle due to accelerated aging of solid insulation in transformers, motors, and capacitor banks. It is usual to regard the dissipation of heat in electrical networks as the product, I^2R, evocative of electrical

FIGURE 9.1 Harmonic current distribution from a nonlinear load affecting an adjacent facility.

losses. For the case of harmonic distortion, the total losses can be expressed as the summation of the individual losses at every harmonic frequency:

$$I^2 Z = I^2_{60\ Hz} Z_{60\ Hz} + I^2_{300\ Hz} Z_{300\ Hz} + I^2_{420\ Hz} Z_{420\ Hz} \qquad (9.1)$$

Because most electrical equipment is specified based on 50-/60-Hz parameters, the addition of harmonic losses can limit the ability of the equipment to work up to the rated value. Harmonic losses are related to the additional heat developed during the operation of nonlinear loads. Putting this in a simple perspective, harmonics losses can be regarded as the difference in heat dissipation between two parallel loads of the same size, one linear and the other nonlinear, when they are fed off from the same source.

Figure 9.1 illustrates this concept. I_h is the harmonic current that produces the additional losses. There is a linear load at the left of the figure and a nonlinear load to the right. The nonlinear load is a variable frequency drive, symbolized here, for simplicity, just as a thyristor. The output of the drive is an AC voltage that, in the example, is stepped up to compensate the voltage drop on a long cable feeding a large load. As previously discussed in Chapter 2, variable frequency drives distort the source side voltage waveform due to the way in which they draw current in slices during the commutation process. The decomposition of the current waveform in Fourier series provides the spectrum of harmonic currents that can be used to calculate the combined contribution of every harmonic component to the total losses. Harmonic losses will generally show up in the form of copper and core losses.

Here it is important to remember that harmonic currents increase the RMS or total effective load current as follows:

$$I_{RMS} = \sqrt{I_1^2 + \sum_{h=2}^{n} I_h^2} \qquad (9.2)$$

Fundamentals of Power Losses in Harmonic Environments

If this current is used in the general equations that depict ohmic losses, the result will describe the losses contributed by all individual harmonic currents. The example in Figure 9.1 is intended to illustrate in a simple way how harmonic loss dissipation may take place in the context of two similarly sized but different types of loads. Both of them draw similar power frequency currents that will produce identical heat dissipation in their feeding transformers. However, the increased RMS current value due to harmonics from the nonlinear load will create added transformer losses in a different fashion as follows.

If there are harmonic filters tuned to provide low-path impedances for all characteristic harmonics of the converter, the several branches of filter, F_h, will ideally absorb them all. In this way, the transformer connected to the nonlinear load is subject to additional losses.

If the filter is not there, harmonic currents will freely propagate upstream, finding an additional path toward the adjacent facility and towards the AC source, as indicated by the dotted lines in Figure 9.1. This will now create added losses on the two service transformers. In these circumstances, the two customer facilities are affected by additional losses arising from harmonic currents created at one of the customer premises.

Assuming the most likely scenario — that the branches of harmonic filter, F_h, are tuned to lower order harmonics, the remaining spectral components, especially those of a higher order, may find a likely path towards the capacitor bank on the adjacent facility. Increased feeder and source impedance losses would need to be accounted for in the calculation of total increased losses. How exactly losses will distribute on the different components can only be determined through a detailed representation of the distribution system elements.

In any case, it is obvious that the transformer on the nonlinear load will be subject to the largest losses, regardless of the existence of the harmonic filter. Specially designed K-type transformers discussed in Chapter 4 are used in these applications; they are expected to dissipate the added energy loss adequately without increasing the transformer temperature beyond design limits. The question that arises here concerns which parameters play a role in the generation of harmonic losses. This is the topic of the next section.

9.3 RELEVANT ASPECTS OF LOSSES IN POWER APPARATUS AND DISTRIBUTION SYSTEMS

Under purely sinusoidal conditions, the calculation of losses in a power system is straightforward because it is based in conventional power flow studies that assume linear impedances throughout the system. The increasing waveform distortion in power systems due to the proliferation of nonlinear loads requires losses to be calculated using more suitable techniques. These involve time series in which voltage and current quantities are expressed comprising the most relevant frequency components other than the fundamental frequency of the system.

As a result, the active, reactive, and apparent power must be determined using the expressions presented in Chapter 1 as follows:

$$P = \frac{1}{T}\int_0^T p(t)dt = \sum_{h=1}^{\infty} V_h I_h \cos(\theta_h - \delta_h) = \sum_{h=1}^{\infty} P_h \qquad (9.3)$$

$$Q = \frac{1}{T}\int_0^T q(t)dt = \sum_{h=1}^{\infty} V_h I_h \sin(\theta_h - \delta_h) = \sum_{h=1}^{\infty} Q_h \qquad (9.4)$$

$$S = V_{rms} I_{rms} \qquad (9.5)$$

$$S^2 = P^2 + Q^2 + D^2 \qquad (9.6)$$

9.4 HARMONIC LOSSES IN EQUIPMENT

9.4.1 Resistive Elements

If we assume that a 1-ohm resistive element is the path for a fundamental current, $I_1 = 1$ A, containing additional third, fifth, and seventh harmonic levels whose amplitudes are inversely proportional to their harmonic order, the RMS current can be calculated as follows:

$$I_{RMS} = \sqrt{\sum_{h=1}^{\infty} I_h^2} \qquad (9.7)$$

$$= \sqrt{1^2 + 0.33^2 + 0.2^2 + 0.1^2} \approx 1.0765 A$$

This small increase in current above 1 A will produce increased losses, ΔP, relative to the case in which current contained no harmonics equal to:

$$\Delta P = (I_{RMS}^2)(R) - (I_1)^2(R) = (I_{RMS}^2 - I_1^2)R \qquad (9.8)$$

$$= [(1.0765)^2 - (1)^2](1) = 0.1588 \text{ W}$$

or

$$\frac{0.1588}{1}(100)$$

This is practically 16% above the case with no harmonics.

Fundamentals of Power Losses in Harmonic Environments

The corresponding THD_I is given by Equation (1.40) in Chapter 1 as follows:

$$THD_I = \frac{\sqrt{\sum_{h=2}^{\infty} I_h^2}}{I_1} \qquad (9.9)$$

$$THD_I = \frac{\sqrt{\sum_{h=2}^{\infty}(0.33^2 + 0.2^2 + 0.1^2)}}{1} = 0.399 = 39.9\%$$

If the assumed current is 2 A with harmonic currents keeping the same proportion relative to the fundamental current, rms current and losses become:

$$I_{rms} = \sqrt{(2^2 + 0.66^2 + 0.4^2 + 0.2^2)} \approx 2.153 A$$

and

$$\Delta P = (I_{rms}^2 - I_1^2) R = [(2.153)^2 - (2)^2] (1) = 0.6356 \text{ W}$$

or

$$\frac{0.6356}{2}(100)$$

This is practically 32% higher than losses with no harmonics.

If we make $P1$ and $P2$ stand for the losses at 1 and 2 A, respectively, their ratio can be expressed as:

$$\frac{P_2}{P_1} = \frac{2.153^2 (1)}{1.0765^2 (1)} = 4$$

Thus, from this example, we observe that the loss increase is proportional to current even under distorted conditions and that the total dissipated power on the resistor increases proportionally to the square of current; thus, if current doubles, losses quadruple.

The corresponding THD_I is the same as for the previous case because the harmonic currents assumed were increased in the same proportion as the fundamental current. From Equation (9.8):

$$THD_I = \frac{\sqrt{\sum_{h=2}^{\infty}(0.66^2 + 0.4^2 + 0.2^2)}}{2} = 0.399 = 39.9\%$$

From Equation (9.6) and Equation (9.8), we can determine the relationship between the rms current and the total harmonic distortion of the current:

$$THD_I = \frac{\sqrt{\sum_{h=2}^{\infty} I_h^2}}{I_1}$$

$$I_{RMS} = \sqrt{\sum_{h=1}^{\infty} I_h^2} = \sqrt{I_1^2 + \sum_{h=2}^{\infty} I_h^2}$$

$$THD_I^2 = \frac{\sum_{h=2}^{\infty} I_h^2}{I_1^2} = \frac{I_{RMS}^2 - I_1^2}{I_1^2}$$

$$(THD_I^2)(I_1^2) + I_1^2 = I_{RMS}^2$$

$$I_{RMS} = \sqrt{I_1^2 (THD_I^2 + 1)}$$

$$I_{RMS} = I_1 \sqrt{THD_I^2 + 1} \qquad (9.10)$$

9.4.2 Transformers

Transformer losses have two components: copper and core losses. Copper losses occur in the windings and are a function of 60-Hz resistance; at increased frequencies, resistance is even increased due to skin effect. Several methods to estimate the additional heating expected from nonsinusoidal loads are discussed in the next sections.

9.4.2.1 Crest Factor

This is the simplest way to express the relation between the maximum and the effective value of a voltage signal and yields $\sqrt{2}$ for the case when the signal is a pure sinusoidal waveform. This ratio is exposed to changing under harmonic distortion of the voltage signal created by nonlinear loads. It was popular in 1988[1] to

Fundamentals of Power Losses in Harmonic Environments

express the impact of harmonics on the voltage fed to computer equipment. The mathematical definition of crest factor is the peak magnitude of the current waveform divided by its true rms value:

$$CrestFactor = \frac{Vpeak}{V_{RMS}} \qquad (9.11)$$

9.4.2.2 Harmonic Factor or Percent of Total Harmonic Distortion

This is the frequently cited total harmonic distortion (THD) factor expressed in Equation (9.9). THD factor measures the contribution of the additional rms harmonic current to the nominal rms fundamental current; however, similar to crest factor, it does not have a means to consider harmonic heating losses.

9.4.2.3 K Factor

The calculation of K factor considers the important effect that frequency has on transformer loss estimation. This factor is defined as the sum of the squares of the harmonic current in p.u. times the square of the harmonic number. In form of equation:

$$K = \sum_{h=1}^{\infty}(I_h^2 * h^2) \qquad (9.12)$$

Alternatively, it can also be expressed as:

$$K = \frac{\sum_{h=1}^{\infty} h^2 I_h^2}{\sum_{h=1}^{\infty} I_h^2} \qquad (9.13)$$

where h is the harmonic order and I_h is the harmonic current of order h expressed in p.u. of the fundamental frequency current.

As expressed by Equation (9.12) and Equation (9.13), K factor takes into account the effect of I^2R, which relates to losses, for every harmonic current component. This is a relevant parameter on the assessment of premature aging of transformer windings because dissipated heat in the form of copper and core losses due to spectral components of the current.

Because K factor takes into account the frequency parameter, it is regarded as the most precise method to estimate the harmonic content of nonlinear loads for the specification of distribution transformers. See Chapter 1 for an additional description

TABLE 9.1
Harmonic Content at the PCC of an Industrial Facility

h	I_h	I_h^2	i_h^*	$i_h^2 h^2$	$I_h^2 h^2$
1	1	1	0.9099	0.8279	1
3	0.33	0.1089	0.3003	0.8116	0.9801
5	0.20	0.04	0.1819	0.8272	1
7	0.14	0.0196	0.1274	0.7953	0.9604
9	0.11	0.0121	0.1000	0.8100	0.9801
11	0.09	0.0081	0.0819	0.8116	0.9801
13	0.08	0.0064	0.0728	0.8956	1.0816
15	0.07	0.0049	0.0637	0.9129	1.1025
17	0.06	0.0036	0.0546	0.8615	1.0404
19	0.05	0.0025	0.0455	0.7474	0.9025
21	0.05	0.0025	0.0455	0.9124	1.1025
Σ		1.2086		9.21	11.13

$$* \ i_h = \frac{I_h}{I_{rms}} = \frac{I_h}{(\Sigma I_h^2)^{1/2}}.$$

of K-type transformers. K-factor transformers are constructed so that the higher the K factor is, the higher the harmonic content that they can handle without additional heating will be. K = 1 would be a conventional transformer not fitted for working in harmonic environments at rated power.

Following Underwriter Laboratories' listing of the K4 to K50 transformers aligned with the ANSI Standard C57.110-1986[2], changes to transformer designs were made to minimize losses. Changes considered increasing the primary winding size to better tolerate the circulating triplen harmonics, getting a design with a lower flux density core and insulated parallel transposed secondary-wiring conductors to reduce resistance involved in the skin effect heating. This looked promising to obtain transformer designs with improved thermal dissipation to minimize the additional losses.

K factor is then an index that determines the changes that conventional transformers must undergo so that they can adequately handle the additional iron and copper losses that will be imposed by harmonic currents, particularly when operating at rated power. This is a needed measure to avoid having to derate transformer nominal capacity when installed in harmonic environments.

9.5 EXAMPLE OF DETERMINATION OF K FACTOR

Assume that the harmonic content observed at the PCC in an industrial facility is that shown in the first two columns of Table 9.1. Calculate the K factor using the expressions shown before.

Using the values obtained in Table 9.1, the K factor according to Equation (9.12) yields:

Fundamentals of Power Losses in Harmonic Environments

$$K = \sum_{h=1}^{21} (I_h(p.u.))^2 \cdot h^2 = 9.21$$

Alternatively, using Equation (9.13), K factor results in:

$$K = \frac{\sum_{h=1}^{21} h^2 I_h^2}{\sum_{h=1}^{21} I_h^2} = \frac{11.13}{1.2086} = 9.21$$

As observed, the calculated K factor results are the same using either of the two expressions described in Equation (9.12) and Equation (9.13).

9.6 ROTATING MACHINES

The difference between the synchronous actual speed of an induction motor (speed at which the magnetic field is rotating) and the actual rotor speed is known as slip frequency. The electromagnetic torque varies as a direct function of the slip. This means that a large electromagnetic torque would require of a large slip frequency, ω_{slip}.

Induction machine losses can be estimated as the difference between the power crossing the air gap through the rotor ($T\omega_{sync}$) and the power delivered through rotor to the load ($T\omega_s$)[3]:

$$P_{losses} = T\omega_{sync} - T\omega_m = T\omega_{slip} \qquad (9.14)$$

This suggests that a small slip will minimize induction machine rotor losses.

A study conducted by Fuchs et al.[4] assessed harmonic losses in the stator of an 800-W, 60-Hz, 4-pole, 1738-rmp, 2.35-A phase current and a 220-V induction motor having stator, rotor, and magnetization parameters as follows:

$$R_1^S = 7.0 \ \Omega, \ X_1^S = 8.0 \ \Omega$$

$$R_1^R = 4.65 \ \Omega, \ R_1^R = 7.3 \ \Omega$$

$$X_m = 110 \ \Omega$$

The outcome of this study showed harmonic stator and rotor losses as a percentage of the total stator and rotor losses, as shown in Table 9.2 and Figure 9.2.

In summary, the findings of this study showed:

A significant effect from stator subharmonic losses with decreasing frequency

TABLE 9.2
Stator, Rotor, and Total Harmonic Losses and Additional Temperature Rise in Induction Machines

Total Harmonic Losses for 800-W Induction Motor as Percentage of Total Losses					
Harmonic Order	3	5	7	11	13
Stator	5.6	3.8	3.7	Less than 1%	Less than 1%
Rotor	17.2	9.1	6.7		

Measured Additional Temperature Rise of Stator 2-HP Motor Induction from Positive and Negative Sequence Harmonic Currents				
Harmonic Order	5	7	11	13
	3	2.1	1.8	1.2

Measured Additional Temperature Rise of Stator 2-HP Motor Induction from Positive Sequence Harmonic Currents				
Harmonic Order	5	7	11	13
		3.8		1.8

Measured Additional Temperature Rise of Stator 2-HP Motor Induction from Negative Sequence Harmonic Currents				
Harmonic Order	5	7	11	13
	5.2		2.9	

Source: Adapted from Fuchs, E.F. et al., *Trans. Power Delivery*, 19(4), Oct. 2004.

Rotor losses that are larger for negative sequence harmonics
Increased rotor losses due to sub- and interharmonics for decreasing frequencies below the power system fundamental frequency

Similarly, increased temperatures were measured by Fuchs et al.[4] on a 2-HP squirrel cage, three-phase induction motor. The results, also summarized in Figure 9.2, revealed:

Slightly larger temperature rise due to negative sequence harmonic voltages acting on the rotor
Rapidly increasing temperature rise form sub- and interharmonic voltage components as frequency decreased below the fundamental power system frequency

In the case of motors supplied from variable frequency drives, elevated levels of harmonics can be involved. This is particularly true when motors are operated at low frequencies. The power loss calculation should include in this case:

Fundamentals of Power Losses in Harmonic Environments

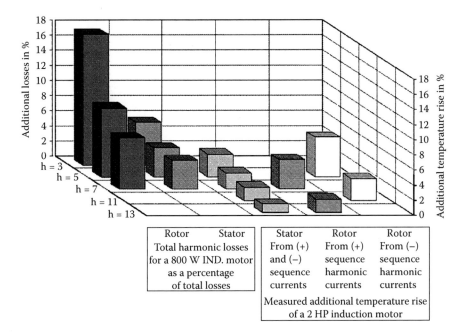

FIGURE 9.2 Harmonic losses and temperature increase in induction machines. (Adapted from Fuchs, E.F. et al., *Trans. Power Delivery*, 19(4), Oct. 2004.)

The power dissipated in the form of losses from the PCC down to the point at which the motor is supplied, including connection cables, transformers, and the variable frequency drive.

The additional power that will be required if the motor is operated at frequencies above 50/60 Hz

REFERENCES

1. Computer Business Equipment Manufacturers' Association (CBEMA), *Three-phase power source overloading caused by small computers and electronic office equipment.* ESC-3 Information Letter, Nov. 1987.
2. IEEE, *Recommended Practice for Establishing Transformer Capability when Supplying Nonsinusodial Load Currents.* ANSI/IEEE C57.110-1986, New York.
3. Mohan, N., Electric drives: an integrative approach, published by MNPERE, 2003. CEI/IEC 1000-2-1:1990, Electromagnetic Compatibility, Part 2: Environment, Sect. 1: description of the environment — electromagnetic environment for low-frequency conducted disturbances and signaling in public power supply systems, first edition, 1990-05.
4. Fuchs, E.F., Roesler, D.J., and Masoum, M.A.S., Are harmonic recommendations according to IEEE and IEC too restrictive? *Trans. Power Delivery*, 19(4), Oct. 2004.

Index

A

Acoustics, field of, 2
AC system, interrelation between, 2m154–157
Active power, 21
Amplitudes, 2, 103
Analysis methods, *see* Harmonic analyses
Apparent power, 21
Arc devices, 35

B

Band-pass filtering techniques, 105–107; *see also* Harmonic filtering techniques
Budeanu in Antoniu, 21
Burst-type harmonic representation, 148–149

C

Cable and equipment, considerations for, 81–83
Cable and line models, currents and, 152–153
Capacitive elements, 4
Capacitor banks, 4, 55, 70–72, 101–102
Capacitor voltage, 109
Circuits, 4
C Message index, 18
Coercive force, 8
Communication interference, 59
Conductors, 83
Correction capacitors, 1
Crest factor, 174–175
CT (current transformers), 90
Currents, harmonic, 44, 65, 150–154
Current transformers (CT), 90
Customer interface, 60

D

Digital simulation, of single-tuned filters, 113–117
Dirichlet conditions, 9
Displacement power factor (DPF), 19
Distortion limits, *see* Harmonic distortion limits, methods to decrease

Distribution static compensator (DSTATCOM), 48
Distribution systems, effects of harmonics on
  abnormal operation of electronic relays, 73
  abnormal operation of solid-state devices, 81
  considerations for cables and equipment, 81–83
  introduction, 69
  lighting devices, 73–74
  miscellaneous effects on capacitor banks, 70–72
  pulsating torques in rotating machines, 74–81
  telephone interference, 74
  thermal effects on rotating machines, 74
  thermal effects on transformers, 69–70
Distribution systems, particulars of, 156–157
Distribution systems, power losses in, 171
DPF (displacement power factor), 19
DSTATCOM (distribution static compensator), 48
Duty cycles, 89

E

Electric furnaces, 55–56
Electric power systems, fundamentals of
  Fourier series, 9–17
  harmonic theory, 2–3
  introduction, 1–2
  linear and nonlinear loads, 3–9
  power quality indices, 17–20
  power quantities under nonsinusoidal situations, 20–24
Electric welders, 1
Electronic ballast, 54
Electronic relays, abnormal operation of, 73
Electrotek Concepts TOP Output Processor, 11
Energy-metering equipment, 83
Equipment, power losses in, 172–176
Equipment, resistive elements in, 172–173
Equipment measurement procedure, 89–90
Even functions, 13

F

Ferromagnetic core equipment, 4

181

Filtering techniques, *see* Harmonic filtering techniques
Filter selection/assessment, 104–105
Fluorescent lamps, 54–55
Fourier series, 9–17
Fuse operation, unexpected, 72

G

Generators, 81–82

H

Half-wave, 134
Harmonic analyses
  analysis methods, 158–160
  examples of, 160–166
  flux of harmonic currents, 150–154
  harmonic propagation facts, 149–150
  harmonic source representation, 142–149
  interrelation between AC system and load parameters, 154–157
  introduction, 139
  power frequency *vs.* harmonic current propagation, 139–142
Harmonic Analysis Program (HASIP), 124–129
Harmonic attenuator elements, series reactors as, 135–136
Harmonic cancellation, multipulse converters and, 47, 134–135
Harmonic current propagation *vs.* power frequency., 139–142
Harmonic currents, 44, 65, 150–154
Harmonic distortion; *see also* Electric power systems, fundamentals of
  limits of, 60–66
  power quality indices under, 17–20
  signature of, 28–29
Harmonic distortion limits, methods to decrease
  cancellation through use of multipulse converters, 134–135
  increase of supply mode stiffness, 132–133
  introduction, 131–132
  network topology reconfiguration, 132
  phase balancing, 136–138
  series reactors, 135–136
Harmonic filtering techniques
  band-pass filters, 105–107
  example: adaptation of a power factor capacitor bank, 110–113
  example: digital simulation of tuned filters, 113–117
  example: high-pass filter at generator terminals, 117–124
  example: using University of Texas HASIP program, 124–129
  general aspects in design of passive filters, 93–94
  introduction, 93
  methodology for design of tuned filters, 108–110
  relevant aspects to consider in design of passive filters, 107–108
  single-tuned filters, 94–105
Harmonic levels, *see* Standardization, of harmonic levels
Harmonic propagation facts, 149–150
Harmonic-related losses, *see* Power losses, fundamentals of
Harmonics measurements
  introduction, 85–86
  measurement procedures, 89–90
  relevant aspects, 90–91
  relevant questions, 86–89
Harmonic source representation, 142–149
Harmonic sources
  electric furnaces, 55–56
  fluorescent lamps, 54–55
  future sources of harmonics, 56–57
  introduction, 27–28
  power converters, 39–54
  rotating machines, 37–39
  signature of harmonic distortion, 28–29
  traditional harmonic sources, 29–56
  transformers, 36–37
Harmonic theory, basics of, 2–3
HASIP (Harmonic Analysis Program), 124–129
Hybrid filters, 93
Hysteresis curve, 8
Hysteresis effect, 36

I

Inductive elements, 4
Industrial arc furnaces, 1
Industrial installations, specifics of, 157
Instantaneous power, 5, 21
Instantaneous voltage, 20
Insulated gate bipolar transistor (IGBT), 6, 42
Integrating equation, 13
Interharmonics, 60, 73
International Electrotechnical Commission (IEC), 59

Index

K

K factor, 19, 175–176

L

Large power converters, 45
Light flicker, 73
Lighting devices, 73–74
Line and cable models, 152–153
Linear/nonlinear loads, 3–9
Load parameters, interrelation between, 154–155
Low-power converters, 46

M

Magnetic flux density/intensity, 7, 36
Measurements, harmonics, *see* Harmonics measurements
Mechanical damping, 81
Mechanical fuses, 74
Medium-sized power converters, 45
Modeling philosophy, 151–152
Motor speed variations, 48
Multipulse converters, 134

N

National Electrical Code (NEC), 19, 83
Network topology reconfiguration, 132
Neutral conducting overloading, 70
Nonlinear loads, 3, 6–9
Nonsinusodal situations, power quantities under, 20–24

O

Odd functions, 13
Ohm's law, 4
Orthogonal functions, 12
Overheating, 6
Overstressing, 70–71
Overvoltage, 148

P

Parallel resonant points, 27, 97–100
Passive filtering techniques, *see* Harmonic filtering techniques
PCC (point of common coupling), 17, 88
Periodic function, 9
Permanent magnetization, 8
Personal computers (PC), 28
Phase angle, 10
Phase balancing, 136–138
Phase-shift transformers, 131
Point of common coupling (PCC), 17, 88
Polarity, 71
Polluted power, 28
Potential transformers (PT), 90
Power converters, 39–54
Power factor, 85
Power frequency *vs.* harmonic current propagation, 139–142
Power losses, fundamentals of equipment, 172–176
　example of determination of K factor, 176–177
　introduction, 169
　meaning of harmonic-related losses, 169–171
　relevant aspects of, 171
　rotating machines, 177–179
Power quality indices, under harmonic distortion, 17–20
Power quantities, under nonsinusoidal situations, 20–24
Power system analysis, *see* Harmonic analyses
Power system measurements, *see* Harmonics measurements
Propagation, natural laws of, 131
Propagation facts, harmonic, 149–150
Proportional integral (PI), 48
Pulsating torques, in rotating machines, 74–81
Pulse width modulation (PWM), 48
PWM (pulse width modulation), 48

R

Reactive power, 21
Remanence, 8
Resistive elements, 4
Resonant conditions, 71–72, 147–148
Root sum square (RSS), 59
Rotating machines; *see also* Harmonic sources
　power losses in, 177–179
　pulsating torques in, 74–81
　thermal effects on, 74
RSS (root sum square), 59

S

SCADA (supervisory control and data acquisition), 85

Series reactors, as harmonic attenuator elements, 135–136
Shaft torque, 78
Single-phase loads, 23
Single-phase *vs.* three-phase modeling, 152
Single-tuned filtering techniques, 27, 94–105
Skin effect, 70
Smart Predictive Line Controller, 56
Solid-state devices, abnormal operation of, 81
Source representation, 142–149
Standardization, of harmonic levels
    harmonic distortion limits, 60–66
    introduction, 59–60
Static power converter, 1
Stiffness, increase of supply mode, 132–133
Subharmonics, 12
Supervisory control and data acquisition (SCADA), 85
Supply mode stiffness, increase supply of, 132–133

T

Telephone influence factor (TIF), 18, 105
Telephone interference, 74
THD (total harmonic distortion), 17, 59, 105, 175
Thermal effects
    on rotating machines, 74
    on transformers, 69–70
Three-phase system, 22–23
Three-phase *vs.* single phase modeling, 152
Total demand distortion, 17
Total harmonic distortion (THD), 17, 59, 105, 175
Total power factor, 19
Transducers, 90
Transformers, 1, 36–37, 69–70, 153, 160, 174

Tuned harmonic filters, methodology for design of, 108–110; *see also* Harmonic filtering techniques
Tuning reactor, 104
Type-k transformers, 131

U

Unbalance detection, 104
Underwriter's Laboratories (UL), 19
Unified magnetic equivalent circuit transformer (UMEC), 36
Uninterruptable power supplies (UPS), 28, 46
University of Texas (HASIP) program, 124–129
UPS (uninterruptable power supplies), 28, 46

V

Variable frequency drives (VFD), 38, 47
VFD (variable frequency drives), 38, 47
Voltage control, 7
Voltage distortion, 60
Voltage drop, 133
Voltage related parameters, 20
Voltage source inverter (VSI), 48
Voltage waveform, 60

W

Watts, 142
Waveform distortion, 1, 2, 86–87, 139
Waveform symmetry, effect of, 14

Z

Zero sequence, 42, 69, 96